Learning Resource Centre

Sustainable Living

Editor: Tracy Biram

Volume 368

Independence Educational Publishers

First published by Independence Educational Publishers

The Studio, High Green

Great Shelford

Cambridge CB22 5EG

England

© Independence 2020

Copyright

Photocopy licence

ISBN-13: 978 1 86168 825 5

Printed in Great Britain

Zenith Print Group

Contents

Introduction

Sustainable Living is Volume 368 in the **ISSUES** series. The aim of the series is to offer current, diverse information about important issues in our world, from a UK perspective.

ABOUT SUSTAINABLE LIVING

The majority of people are concerned about how our demand for resources is affecting the planet. Our throw-away society must change in order for us to preserve the world we live in. This book looks at issues and solutions, how to keep the environment healthy and minimise our impact, and how we can live more sustainably. It also considers sustainable food and fashion and how we can be more responsible in our choices.

OUR SOURCES

Titles in the **ISSUES** series are designed to function as educational resource books, providing a balanced overview of a specific subject.

The information in our books is comprised of facts, articles and opinions from many different sources, including:

- Newspaper reports and opinion pieces
- Website factsheets
- Magazine and journal articles
- Statistics and surveys
- Government reports
- Literature from special interest groups.

A NOTE ON CRITICAL EVALUATION

Because the information reprinted here is from a number of different sources, readers should bear in mind the origin of the text and whether the source is likely to have a particular bias when presenting information (or when conducting their research). It is hoped that, as you read about the many aspects of the issues explored in this book, you will critically evaluate the information presented.

It is important that you decide whether you are being presented with facts or opinions. Does the writer give a biased or unbiased report? If an opinion is being expressed, do you agree with the writer? Is there potential bias to the 'facts' or statistics behind an article?

ASSIGNMENTS

In the back of this book, you will find a selection of assignments designed to help you engage with the articles you have been reading and to explore your own opinions. Some tasks will take longer than others and there is a mixture of design, writing and research-based activities that you can complete alone or in a group.

FURTHER RESEARCH

At the end of each article we have listed its source and a website that you can visit if you would like to conduct your own research. Please remember to critically evaluate any sources that you consult and consider whether the information you are viewing is accurate and unbiased.

Useful Websites

www.commonslibrary.parliament.uk

www.edie.net

www.eufic.org

www.euronews.com

www.fairtrade.org.uk

www.independent.co.uk

www.inews.co.uk

www.learnervegan.com

www.onehome.org.uk

www.openaccessgovernment.org

www.planetschooling.com

www.shoutoutuk.org

www.studentnewspaper.org

www.telegraph.co.uk

www.theconversation.com

www.theecologist.org

www.theguardian.com

www.yougov.co.uk

www.youmatter.world

Chapter 1

Sustainability

Sustainability – what is it? Definition, principles and examples

What is sustainability? What does it mean? What are the principles and pillars behind sustainability? What examples of sustainability are there in technology, food, workplace, business or transportation areas? How does sustainability relate to demand and supply? Find the answers to these and further questions below.

What is sustainability?

There is no universally agreed definition of sustainability. In fact, there are many different viewpoints on this concept and on how it can be achieved.

Etymologically, the word **sustainability** comes from **sustainable** + **ity**. And **sustainable** is, for instance, a composition of **sustain** + **able**. So if we start from the beginning, to **'sustain'** means 'give support to', 'to hold up', 'to bear' or to 'keep up'.

What is sustainability, then? Sustainable is an adjective for something that is able to be sustained, i.e, something that is 'bearable' and 'capable of being continued at a certain level'. In the end, sustainability can perhaps be seen as the process(es) by which something is kept at a certain level.

Nonetheless, nowadays, because of the environmental and social problems society is facing, sustainability is commonly used in a specific way. Therefore, sustainability can be defined as the processes and actions through which humankind avoids the depletion of natural resources (which is influenced by the way societies are organized) to keep an ecological balance so that society's quality of life doesn't decrease.

In this way, we can say that resources exploitation, manufacturing operations, the direction of investments, technological developments, wealth distribution and institutional changes, among others, are being sustainable if they don't hurt the ecosystem services and if they allow for society's quality of life not to decrease.

Definition of sustainability and sustainable development: what's the difference?

The views on sustainability seem to have a stronger focus on the present moment and on keeping things above a certain level. By its turn, sustainable development focuses more on a long-term vision. In fact, sustainable development has a universally agreed definition that was first written in the Brundtland Report.

By adding the concept of 'development', sustainable development means that humankind should satisfy its current needs without compromising the ability of future generations to do the same. Along with it also comes an idea of societal progress and an increase in quality of life.

That's with an agenda for 2030 where 17 sustainable goals were adopted by the UN members in NY in 2015. Among them are goals such as ending poverty and hunger, ensuring good health and well-being for all, providing quality education and achieving gender equality.

Principles of sustainability: the 3 pillars of sustainability

What is sustainability? The principles of sustainability are the foundations of what this concept represents. Therefore, sustainability is made up of three pillars: economic, social and environmental. These principles are also informally used as profit, people and planet.

John Elkington, from a sustainability consultancy firm, was one of the first people to integrate these three principles. He argued companies needed to start considering this triple bottom line so that they could help the world thrive in the long run.

At the same time, consumers and citizens unsatisfied with the long-term damage (both on wealth distribution and the environment) caused by the corporate focus on short-term profits, turned sustainability into a mainstream concept able to ruin a company's reputation and profits. Today, sustainability is often spoken of with regard to climate change, which threatens life as we know it and is largely caused by industrial practices. That's one of the reasons why today many companies have corporate responsibility (CRS) strategies.

Some news channels are sharing the positive effects the new coronavirus outbreak is having on the environment and climate. However, unfortunately, coronavirus is likely to be bad news for ecology in the long-term because it is tied to a dysfunctional economic system. We explain why, here.

Examples of sustainability: a long-term vision

Sustainability encourages people, politics, and businesses to make decisions based on the long term. In this way, acting sustainably encompasses a temporal framework of decades (instead of a few months or years) and considers more than the profit or loss involved.

Let's find out about different examples of sustainability depending on the industry:

Technology: examples of sustainability in technology

The use of electronic devices is growing every day. Nonetheless, these devices are made of Earth minerals extracted by the mining industry. Mining can be a very polluting industry and the development of new sites certainly has an impact on deforestation.

Therefore, being sustainable in the tech field means using your devices for a long period instead of constantly replacing them with the latest upgrades. It is also about making sure you dispose of old devices in a responsible way.

Soon, sustainability in technology will also be about how the (mostly) lithium-ion batteries of electric cars and solar panels will be disposed of. Companies focusing on recycling these batteries and building products whose core can be maintained and batteries easily replaced with recycled ones, will be at the forefront of sustainable tech.

Fashion: examples of sustainability in fashion

Fashion, especially fast fashion, focuses on speed and low cost to deliver frequent new collections. Nonetheless, the problem with this industry is its negative environmental impact. On one hand, it uses toxic chemicals that cause water pollution and which may contaminate soil too, if wrongly disposed of.

On the other hand, there's a lot of textile waste and many clothes are made of synthetic fibres which, while being washed, escape to the sea in the shape of microplastics. If a company makes clothes with resistant materials, uses sustainably produced cotton, applies principles of circular economy in its value chain and uses fewer toxic chemicals, it is being responsible with the environment.

At the same time, sustainability is also about being socially responsible. And overall, the fashion industry has a poor record in this area. If you pay attention most labels show that clothes are being made in distant places such as China, Bangladesh or Vietnam.

Apart from the pollution of transporting these items, the manpower behind the manufacturing of these clothes is what's most worrying. People in these countries usually get really low wages and work under bad conditions. They can hardly improve their social situation and mostly keep

Three pillars of sustainability

FASHION · EMPLOYMENT · FOOD

on working just to pay the bills and survive. This largely contributes to the inequality we see in the world since in 2018 the rich got richer and the poor poorer, according to Oxfam's latest report.

Transportation: what is sustainability in transportation?

A report from the IPPC says 14% of all greenhouse gas emissions come from transport and are mostly due to passenger cars. Contrary to what many believe, planes, cargo ships or even trucks aren't the main contributors to CO2 emissions and cars can assume much of the blame. So unless someone is driving a car with 4 or 5 passengers, taking public transportation, especially trains and buses are more sustainable choices. And if one can simply walk or cycle it'd be even better.

Today, there are even more sophisticated solutions to reduce the pollution caused by moving around. At a vehicle level, the popularity and development of alternatives like electric cars (or even hydrogen cars) or electric scooters are growing at a high rate. At the same time, solutions like carpooling, where drivers share their cars and save some money (and pollution) are great alternatives. Not to mention the fact that more companies are letting their employees work remotely or from home, saving the number of kilometres travelled, too.

Zero waste as an example of sustainability

The zero-waste movement is a lifestyle that encourages people to use all types of resources in a circular way, just like the natural world does. The ultimate goal of this philosophy is to avoid resources following a linear route and ending up as trash in the oceans or as landfill. For this, people must refuse what they don't need, reduce what they're getting, reuse it and recycle or compost it.

Linked with this lifestyle is also a minimalist way of living, where people are often invited to leave behind and refuse what they don't need. The movement is also very well known by people taking their own bags to shops to bulk-buy commodities such as chickpeas, rice or liquid soap. The goal is clear: not to take any trash home.

The food industry: examples of sustainability in this area

A company that tries to grow its crops by not using (or using few) toxic pesticides and focuses on organic farming and biomimicry practices is certainly a less polluting one. If it pays fair wages to its employees and manages to still be competitive on the market, it is also being responsible when it comes to profit, people and planet.

Workplace: examples of what sustainability in the workplace is

The workplace can also be organized in a sustainable way. For instance, companies betting on new technologies and becoming paperless or that provide conditions and training for employees to recycle are being careful about waste management.

At the same time, only using air-conditioners for very extreme temperatures (large energy waste and GHG emissions), opening the blinds when there's sunlight and avoiding plastic cutlery are also good ways to have a sustainable workplace.

Operations and value-chain: where's the sustainability?

Let's analyse sustainability in operations by imagining a company with very high energy costs, such as a steel manufacturers. If it is economically viable, the company could install solar panels and power its operations with this energy. It'd be a medium-long term investment that could be economically positive in the long run.

At the same time, the company would be using renewable energy, which is especially important in places where the electricity grid works mostly on fossil fuels.

A company's strategy (CSR): where is its sustainability?

A company's corporate social responsibility (CSR) is a strategy that integrates the policies and practices of firms wanting to create value on their triple bottom line (people, planet, profit). So besides taking care of their workplaces and trying to be eco-friendly along their value-chain, companies with a sustainability mindset are also concerned about social issues like gender equality, happiness at the workplace and taking care of the communities affected by their activities.

At the same time, they don't underestimate the financial side of the business, where profit is a basic condition for organizations to survive – yet, it's not the main reason or the main purpose why these businesses exist.

Sustainable cities: what does it mean to be a sustainable city?

Sustainable cities are cities that have strong social, economic and environmental performances. They have good scores when it comes to air pollution, availability of public transportation, the number of educated and employed people, the percentage of green spaces, energy consumption and access to drinking water.

Presumably, sustainable cities would be better prepared to face the challenges of urban areas as society develops and as climate change events get more frequent and intense.

Waste management: is there sustainability in waste management?

A factory that takes proper care of its industrial waste and doesn't drop it in a nearby river or land is acting in a sustainable way. In fact, this factory is being responsible for avoiding the short-term costs of damaging disposal that could cause expensive and impactful long-term environmental damage.

At the same time, companies looking for less polluting packaging alternatives are also good sustainability models to follow. Since plastics are polluting land and seas and harming

ecosystems and biodiversity, it's a good idea that businesses invest in new designs that allow products to be more resistant and even re-manufactured. On top of this, if biodegradable materials are being used, even better.

The connection between supply, demand, and sustainability

Supply and demand. Demand and supply. We often hear of these two concepts and it's not hard to think about their connection with sustainability and sustainable development. Supply and demand are economic forces of the free market that control what suppliers are willing to manufacture and what consumers are willing to purchase.

Specifically, supply means how much of a certain product, commodity or service suppliers are willing to 'give' or produce at a certain price. And demand refers to how much of this product or service consumers are willing to purchase at a particular price.

The relationship between demand and supply carries the forces behind the allocation of resources. According to market economy theories, demand and supply theory will allocate resources in the most efficient way. The connection between this theory and sustainability is that nowadays we're going over Earth's biocapacity because we're 'demanding too much'.

This demand is happening not only because the population is increasing. It is also due to the appealing equilibrium price which is, among other things, influenced by the law of mass production.

At the same time, sustainability is often spoken of in terms of the supply chain. In this case, it means that companies should be concerned about the sustainability of their suppliers' processes.

What is the ecosystem's sustainability?

The sustainability of ecosystems is about keeping the ecological services working. This means an ecosystem's footprint cannot exceed its biocapacity. But what is biocapacity?

Biocapacity definition: what is biocapacity?

The definition of biocapacity, according to WWF, is 'the capacity of ecosystems to produce useful biological materials and to absorb waste materials generated by humans, using current management schemes and extraction technologies.'

Also, according to the Global Footprint Network, biocapacity can change because of climate and depending on which ecosystem services are considered useful inputs to be used in the human economy. Also, according to the National Footprint Accounts, 'the biocapacity of an area is calculated by multiplying the actual physical area by the yield factor and the appropriate equivalence factor. Biocapacity is usually expressed in global hectares'.

6 May 2020

Are the UK's young more sustainable than gen X and baby boomers?

Are millennials sustainable? That is the question.

Millennials and their even younger counterparts in Generation Z have an unearned reputation as being overly materialistic consumers, but why and how? Millennials might be more likely to travel than older and younger generations but given that many of these travellers are doing so on a budget, it shouldn't be perceived as an extravagance. Of course, plane travel arguably creates a large amount of pollution, but there are other ways these young nomads might be off setting their carbon footprints.

In fact, millennials (those currently aged between 22 and 35, depending on who you ask) are actually far more likely to shop ethically, eat more locally-sourced food and adopt a vegan diet. Indeed, according to Fiona Dyer, a consumer analyst at Global Data, 'the shift toward plant-based foods is being driven by millennials, who are most likely to consider the food source, animal welfare issues, and environmental impacts when making their purchasing decisions.

This rise of green consumerism amongst the younger generations is indicative of a wider shift in perceptions towards what it means to 'live green'. Veganism is very en-vogue right now, with January 2019 seeing a significant increase in the annual 'Veganuary' movement. Plastic straws are also being slowly phased out by countless major restaurant chains, which chimes neatly with the fact that half of all digital consumers say that environmental concerns directly impact their purchases.

When asked if they would pay more for eco-friendly products, 61 per cent of millennials also say they would willingly do so, whilst only 46 per cent of baby boomers said likewise. Many older millennials are even starting to grow their own veg at home and live off the land on a completely plant-based diet. This involves investing in polytunnels and irrigation systems that require a lot of knowledge to install and operate.

Eco-friendly products are not only better for the environment but they are generally better for your health too, and with an increasingly health-obsessed younger generation, that might also explain why they are jumping on board the eco bandwagon in greater numbers and with greater voracity. This surely shatters the illusion that millennials are lazy or refuse to be sustainable?

So, if younger generations are more ethically minded, why do they have this reputation of being entitled and wasteful? Which is simply not true. It might have something to do with the 'throwaway culture' that has developed over the past few decades, but the truth is that tech-savvy young consumers are that much more likely to invest in sustainable products and sustainable brands.

Currently, millennials make up around 25 per cent of the world's population and 84 per cent of them believe it is their responsibility to change the world for the better by living greener and more responsibly. This is the kind of generation we need if the generations that follow them are to stand a real chance at making it to the next millennium.

3 June 2019

Sustainable Development Goals: how is the UK performing?

I n 2015 the UK Government committed to achieving the UN's Sustainable Development Goals (SDGs). These goals aim to improve peace, prosperity, access to healthcare and education and tackle climate change worldwide.

The SDGs, also known as the 'Global Goals', replaced the 2001 Millennium Development Goals, which only covered developing countries. The new goals apply to advanced economies like the UK.

Each country signed up to a deadline of 2030 to achieve the SDGs and has to report on how they are performing.

On 26 June, the UK Government published its report on the implementation of the SDGs so far. The UK's report – the Voluntary National Review (VNR) – will be presented at the High-Level Political Forum on Sustainable Development scheduled for 16-18 July in New York.

Critics argue that the UK's performance has been inadequate in important policy areas, including combating hunger and food insecurity at home. They also state the Government has not established effective structures and processes for implementing the goals. There has also been dissatisfaction over how the Government conducted the VNR process.

Public awareness of the SDGs appears still to be low. Only 9% of respondents to a 2018 'Aid Attitudes' survey knew what the SDGs were.

This insight looks at how the UK is performing so far.

What are the SDGs?

There are 17 goals, with 169 associated targets for human development (plus 230 statistical indicators). The rallying cry is 'leave no one behind'.

The UK's performance at home

In the VNR, the UK Government claimed these achievements:

◆ a high-quality health service, free for all at the point of use;

◆ high and rising standards of education;

◆ increasing employment, with more women and disabled people in work;

◆ progress made on climate and the environment; and

◆ some of the world's strongest legislation on equality issues.

At the same time, the Government acknowledged areas that need further work, including:

◆ tackling injustice to ensure no one is left behind;

◆ further increasing efforts to address climate and environmental issues;

◆ ensuring the housing market works for everybody;

◆ responding to mental health needs; and

◆ supporting a growing and ageing population.

In July 2018 the NGO coalition UK Stakeholders for Sustainable Development (UKSSD) offered a 'score card' on the UK's domestic performance on all seventeen SDGs. Summing up, it said:

Out of 143 relevant targets [...] the UK is performing well on 24% (green), with 57% where there are gaps in policy coverage or performance is not adequate (amber), and 15% where there is little or no policy in place to address the target or the performance is poor (red).

Examples of targets where the UK scored 'red' included: 'recognizing and valuing unpaid care' (part of goal 5 – gender equality) and 'achieving and sustaining income growth of the bottom 60% of the population at a rate higher than the national average' (part of goal 10 – reduced inequalities).

Earlier this year the Environmental Audit Committee published a report criticising the Government's domestic performance in combating hunger and food insecurity.

The UK's performance abroad

In terms of its performance across the rest of the world, the UK Government stressed in the VNR it was meeting the UN's target of spending 0.7% of gross national income (GNI) on aid. The target became a legal obligation in 2015. Since then it has been met each year.

Sustainable development goals

The Government also highlighted tackling climate change, addressing the 'root causes of extreme poverty' and contributing to 'inclusive and sustainable economic growth' as areas of strong performance.

Prior to the publication of the VNR, the UK network for organizations working in international development, BOND, published its own 'goal-by-goal' assessment of the UK's 'global contribution'.

It claimed that UK aid policy was undermining its stated commitment to the SDGs by not investing enough in 'human development for all, gender equality and women's rights, and social equality, in direct collaboration with relevant communities and civil society'.

It also criticised the Government for moving the focus of aid policy away from its 'primary purpose of poverty eradication' since 2015, so reducing the UK's 'potential impact on the SDGs'.

It added that the UK should be doing more to reform the international tax system and support 'domestic resource mobilisation' and on 'climate action and environmental sustainability'.

BOND concluded the UK should be investing more in 'capacity-building and empowerment programmes', deepening and broadening its range of partnerships.

What happens next?

Each goal-specific chapter of the VNR sets out 'key challenges and actionable next steps' for that goal. More broadly, the VNR identifies a 'number of over arching themes that the UK will continue to focus on, including collaboration, data and financing for the goals'.

On collaboration – an issue which critics have expressed concerns about – the Government says:

The UK government will review and further strengthen the existing means and mechanisms to oversee its contribution to domestic delivery of the goals, building on the Single Departmental Plan process [...] an effective mechanism will also be established to enhance stakeholder engagement and cooperation with government in the domestic sphere.

On financing, the Government's priority will be 'harnessing the potential of UK financial services and institutional investors'. This approach could also court further controversy.

Less contentious (unless progress stalls) will be the Government's commitment to continue filling in 'data gaps' on domestic implementation of the goals. The Office for National Statistics is currently able to source data for 74% of the statistical indicators, leaving 26% of them still unmeasured.

Whether these steps will satisfy critics remains to be seen. UKSSD welcomed the proposal to establish a stakeholder engagement mechanism but regretted that there was no explicit commitment to developing a national plan or to the SDGs being coordinated by a 'more appropriate domestic department' in future. BOND asked: 'has the UK missed an opportunity for global leadership?' Whatever their assessment of the UK's performance so far on the SDGs, all may agree with the Government when it says in the VNR's conclusion that there remains a long way to go on the 'journey to 2030'.

4 July 2019

EU set to miss all 17 Sustainable Development Goals

By Sarah George

Almost every nation within the EU is failing to deliver the amount of progress needed for the bloc to meet any of the UN's 17 Sustainable Development Goals (SDGs), with a 'major' lack of progress recorded against SDG 13, Climate Action.

That is according to a damning new report from the UN's Sustainable Development Solutions Network (SDSN) and the Institute for European Environmental Policy (IEEP).

The report measures the 'distance' between each EU member state's current position on environmental and social issues covered by the SDGs and the framework's targets for 2030. It then analyses existing national policy frameworks and business commitments to estimate how rapidly each nation is likely to progress on each Global Goal in the coming decade.

While emphasising that many European nations are global leaders on SDG progress – with Denmark, Sweden and Finland among the world's closest to achieving the Goals – the report concludes that no EU nation is on track to achieve all 17 Goals by 2030, meaning the bloc is likely to miss all of their over arching ambitions.

Goals where bloc-wide progress has been broadly slow include SDG 13, Climate Action; SDG 14, Life Below Water;

SDG 15, Life on Land; and SDG 12, Responsible Consumption and Production. The EU nations ranked as the most off-track to delivering on these ambitions, and the others detailed in the SDG framework, are Bulgaria, Romania and Cyprus.

The UK, meanwhile, ranked 12th out of a possible 28th. According to the report, 'major challenges' remain in the UK's ability to meet SDGs 12, 13, 14 and 15, with national progress having decreased in the past year. Indeed, the only Global Goal which the UK is on track to meet and where progress has increased quickly in recent months is stated to be SDG 9, Industry, Innovation and Infrastructure.

Policy overhaul

In order for the EU's SDG progress to be put on track, the report recommends that the EU and its member states undertake a series of policy 'transformations', aimed at addressing not only the bloc's negative SDG impacts on a domestic basis, but also its 'spillovers' – the ways in which it is hampering progress in other nations through the impact of sectors such as international transport and sourcing goods such as seafood and forestry products.

Its over arching call to action is for a European Green New Deal including measures to completely decarbonise energy by 2050, transform land use and food systems in line with the Paris Agreement and make the circular economy a reality.

European Commission President-elect Ursula von der Leyen recently said she wanted the Green New Deal to become her Administration's 'hallmark'. Her draft version of the policy, which is due to be enshrined in law within the first 100 days of the new Commission, includes boosting the EU's 2030 emissions target from a 40% reduction to 50% and setting a 2050 net-zero target.

Other recommendations from the IEEP and SDSN include more investment in low-carbon power and transport infrastructure; further funding for education and innovation; and policies preventing member states from creating negative environmental and social impacts overseas.

Ultimately, the two bodies believe the EU must place the SDGs at the centre of its diplomatic, development, and collaborative work in order to deliver joined-up progress.

'By creating a new, inclusive low-carbon circular industry and agriculture, Europe can show the world that it is possible to preserve economic prosperity, while at the same time reducing inequalities and protecting the natural resources that we all depend on for our health and nutrition,' the IEEP's executive director Celine Charveriat said.

22 November 2019

www.edie.net

Salford beats Brighton and Bristol to title of 'greenest place to live'

The former 'dirty old town' has more energy-efficient homes, more green spaces, more recycling and lowest CO2 emissions, says study.

By Tom Wall

Salford may have been fondly dubbed a 'dirty old town' by folk singer Ewan MacColl and depicted as full of smoky chimneys by LS Lowry, but new research has crowned it the greenest place to live in England and Wales.

The city, part of Greater Manchester, is more sustainable than places such as Brighton, where Caroline Lucas is Britain's only Green party MP, and Bristol, a former European Green Capital, according to a study to be released this week by the Centre for Thriving Places.

In this socialist stronghold of the north-west – represented by the Momentum-backed Labour leadership hopeful Rebecca Long-Bailey – the leftwing council, run by Labour mayor Paul Dennett, is building the most energy-efficient new homes in England and Wales and preserving and creating more green space than any other council.

The city also has lower-than-average CO2 emissions, lower-than-average energy consumption levels and higher-than-average recycling rates.

'There have been a lot of innovative things going on in the north-west in terms of participatory democracy, community wealth-building and the environment,' said Liz Zeidler, chief executive of the Centre for Thriving Places. 'Places such as Salford seem to be quite cleverly absorbing these ideas. Its leaders have been willing to try something new and now they are starting to see the results.'

Dennett said Salford had invested millions in green infrastructure and had embraced the decarbonisation agenda long before it was in vogue. 'Our carbon management programme was put in place in 2009 and has helped to save 14,000 tonnes of CO2,' he said. 'And 60% of Salford is now green space.'

According to the study, the least sustainable area in England and Wales is Stockton-on-Tees in the north-east. It has well over double the average CO2 emissions, nearly half the average recycling rates, and fewer trees than most authorities.

The full analysis – which rates 363 local authorities on 60 indicators covering sustainability, equality and local conditions – reveals that Richmond upon Thames offers the best quality of life overall. Residents of the affluent south London borough have the best mental and physical health, along with the most trusting neighbours.

Stockton-on-Tees also has one of the worst scores overall. Its poorest children make less progress than those living in most other areas, while the scale of its health challenges dwarfs those found elsewhere. It has the largest life-expectancy gap in England and Wales, with people in deprived areas dying on average more than 16 years earlier than in the better-off parts of the borough.

One of the most culturally diverse areas in the study, Brent in north-west London, was found to be the most egalitarian place in England and Wales. Its council chamber is fully representative of its black and minority ethic population – an achievement matched by only 21 other councils. It has more than half the average gender pay gap, with children from disadvantaged backgrounds standing a better chance of doing well at school and getting good jobs than those in 80% of areas in England.

Brent said its good schools enabled equality of opportunity. 'The other secret to our success is that we know that diversity is our greatest strength, so we proudly celebrate it,' said its deputy council leader, Margaret McLennan.

Stockton-on-Tees said it accepted it had problems, but the study did not reflect the area's many strengths. 'We're the economic powerhouse of the Tees Valley, contributing about a third of its economic output, and we're home to internationally recognised brands,' said Bob Cook, leader of the council. 'But we do, of course, have our challenges. We have significant areas of deprivation and serious health inequalities.'

He added that the area's energy-intensive industries were meeting government emissions targets, and its weekly bin collections were popular with residents.

14 March 2020

The 10 R's of sustainable living

By Lucy Legan

Exciting times ahead. Climate change is now a part of the mainstream media. Environmental groups agree, it's now or never! And there's thousands of people ready to change their mindset to become more sustainable. The question that we are often asked is 'where do I start?' So here are a few guidelines to get you started on your journey – The 10 R's of Sustainable Living.

Before buying new objects run the 10 R's through your head. If you can't remember them off hand, write a little note to yourself on your telephone or put them on your fridge. The 10 R's remind us to reflect on our actions, helping us to walk our talk.

Respect, Responsibility, Refuse, Reduce, Rethink, Repurpose, Reuse, Repair, Recycle and Restore

Respect

We belong to the earth. And as earthlings, we can learn something from our source and the support – Planet Earth. Nature is the source of the creativity called 'life'. We are not separated from the earth. This philosophy is based on the first ethic of 'care for the earth', followed by 'care for people', challenging the foundation principles and current ethics of society.

An Earth Ethic will ensure steady improvement in the quality of life for this and future generations that respects our common heritage – the planet on which we live. The challenge lies in a willingness to do things differently than we have in the past. So before you buy anything new ask yourself, 'am I respecting life?'

Responsibility

What do we do with the 27 million wrecked cars every year, or the 25 million television sets which are discarded annually? Is there room for this garbage? And the refrigerators, blenders, vacuum cleaners, dishwashers, microwave ovens etc? Have you stopped to think?

With a little imagination it's possible to have an idea of what could happen if this trend continues. Bunyard and Morgan-Grenville have suggested that if we continue to use the planet's nonrenewable resources, such as oil, coal and other minerals, at current rates of use, and continue to misuse renewable resources such as fertile soil and forests, sometime in the future the entire ecological system may fall apart.

But we're not going there! Take responsibility for your shopping desires. Control it and we won't have to go down the road of destruction. Take it personally!

Refuse

Today, here and now, there are certain things that we should automatically refuse.

Plastic bags – An estimated 500 billion to 1 trillion plastic bags are consumed worldwide each year damaging waterways and harming ocean creatures. Refuse single use plastics. Go shopping with cloth bags and promote the green.

Straws – Over 182.5 billion plastic straws are used per year. Once again hard to dispose of and choking wildlife. Refuse straws.

Disposable coffee cups and lids – In Australia, 1 billion takeaway hot drink cups are thrown away every year. Refuse coffee cups and take your own. The list can go on!

With China and other countries refusing to take our recycling we need to reduce the amount of single use plastic.

Reduce

Reducing the amount of purchased goods is an efficient and sensible measure to save energy and money. For example, do you really need to buy a clothes dryer or can you dry your clothes in the sun?

Fast fashion is becoming a huge problem. Australians buy an average of 27 kilograms of new textiles each year and then discard about 23 kilograms. Milburn states that 'Most of our textiles are just buried in landfill, so we can continue consuming without guilt. In a finite world, we can't keep pretending this doesn't matter. We all have to be accountable for our waste.'

Rethink

Think about what you buy as this will help the environment and save you money in the long run. Look for better information on durability of products, if replacement parts can be obtained, and how much energy you consume.

Ask yourself if you really need that new item of clothing. Can you buy it second hand, swap or trade it on eBay or Facebook groups? Or can you go without it? Search for sustainable fashion.

Try to buy products with as little packaging as possible. Buying in bulk or in large boxes, rather than individually wrapped products, means less garbage. Buy items that can be used repeatedly (non-disposable).

Repurpose

Upcycling is the new buzz word. It's the process of transforming by-products, waste materials, useless, or unwanted products into new materials or products of better quality or for better environmental value. This toaster was rubbish and now turned into a funky little plant holder.

Reuse

Reusing objects makes sense. There is no reason to succumb to the pressure of buying the latest model, even if it's faster or brighter. Generally, the longer the product lasts, the less impact it causes environmentally.

If an object has broken, try to fix it before you think about buying another.

Repair

In the book *From Mangle to Microwave*, Christina Hardymeny describes that 'in 1922, an enthusiast of new domestic machines estimated that a washing machine could last for 20 years. Today, with all the improvements in washing and rinsing, its average durability is about 5 to 7 years.'

Durable goods such as kitchen equipment, sound equipment, furniture, automobiles and washing machines play an important role in our lives. Our expectation is always that they last many years, but today, if they work for 10 years we will be very lucky. That means more junk. The production and disposal of these products are causing major impacts on the environment.

The main raw materials used to make durable goods are oil and copper, two scarce resources on the planet. To get an idea, only durable electronics discard approximately 65,000 tons of hazardous waste per year!

Recycle

Once the life of the object has ended, it is time to recycle. Swap/trade it as part of paying for a new one. Or sell/swap/trade it to a person who is looking for replacement parts. You can also use creativity and transform one object into another.

Restore

Working with nature is an exciting challenge and we can have a positive impact on our loved ones and the planet.

Take it personally! The ethics of care for the earth, care for people and share surplus are the guiding ethics of permaculture. As earthlings, we can work towards restoring the planet to beautiful forms.

Start small. Your backyard can become a habitat. It involves becoming aware of the nature of your place and taking some action, however small, to create a special place for the abode of mother nature. Plant a garden for pollinators such as butterflies, bees, even bats.

As earthlings we need to be in a supportive partnership with nature creating the possibility of a mutually beneficial situation. This ethical partnership will support a sustainable future for all species.

12 May 2019

Britons say more needs to be done to encourage recycling

More frequent collections, local recycling facilities and cash incentives would encourage Britons to recycle more of their waste.

By Connor Ibbetson, Data Journalist

The British public want supermarkets to do more to reduce plastic packaging, but also think that more needs to be done to make sure products are recycled after use.

The majority (83%) of Britons say that more needs to be done to encourage recycling in the UK, with only 12% saying they think they get all the help they need to make sure their waste is reused.

Among the biggest issues keeping Brits from recycling more in the UK are a lack of local facilities, councils not collecting certain types of items from the kerbside and confusing rules.

One in six Brits (16%) say they would recycle more if their local authority collected more types of recyclables, more often, from their kerbside bins. Another one in nine (11%) want better or more accessible community recycling facilities closer to home.

Despite half (50%) of Brits knowing about the deposit recycle scheme (the scheme, popular in European countries, sees members of the public paid for recycling) only 9% of Brits say that financial compensation through tax reductions or cash for the time they put into sorting and recycling waste would get them recycling more.

One in twenty Brits say they won't recycle anymore until they are assured that items they send for recycling are actually recycled and not destined for the landfill like regular rubbish.

However, 10% of Brits say they think they already recycle all of their waste, and nothing could help them to recycle more.

4 November 2019

www.yougov.co.uk

Brits think more needs to be done to encourage recycling

Do you believe enough is being done to help with and encourage recycling in th UK, or do you think more could be done? (% of 2,065 UK adults)

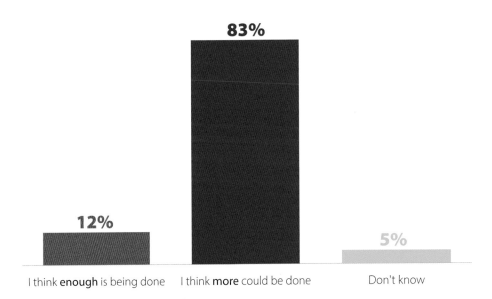

- 83% — I think **more** could be done
- 12% — I think **enough** is being done
- 5% — Don't know

Source: YouGov: 19th - 20th September 2019

Sustainable living for beginners – a starter guide

By Lauren McKnight

I f you've been watching the news, scrolling through Facebook, or keeping up with politics, then you know that climate change has been a hot topic lately. Climate change is threatening our future, and if we don't act now, we could reach a point of no return.

While monumental change may be out of our hands, there are still choices we can make that will reduce our environmental footprint.

If you don't know what it means to live a sustainable lifestyle, or you're just looking to get started and learn more, you are in the right place. In this post, I will explain what sustainable living is, the benefits of sustainable living, and ten simple tips that you can start TODAY to kick-off your 'sustainable living for beginners' journey.

What is sustainable living for beginners?

So what does 'sustainable living for a beginner' mean anyway? Simply put – it is an individual's attempt to reduce their consumption of Earth's natural resources in order to reduce their carbon footprint and save resources for future generations. It could also mean replenishing the resources we have used and producing less waste.

You could extend sustainable living into any area of your life such as:

◆ Transportation/travel

◆ Diet

◆ Energy consumption

◆ Personal resources

The concept of sustainable living is easy – use less of Earth's resources and meet our needs without jeopardising future generations' ability to meet their needs. But it isn't enough to know what sustainable living is – everyone needs to know why it is important.

Why is it important to live a sustainable lifestyle?

The reason living a sustainable lifestyle is important is pretty straightforward – if we don't take care of the environment, we reduce our quality of life here on Earth, natural resources will grow scarce, and animal populations will be in danger of becoming extinct. We risk some pretty grave dangers if we don't make a change, and make it soon.

The good news is that sustainable living is the root of this change, and there are things you can do at home, in your community, and around the world to do your part to protect the planet. Sustainable living for beginners doesn't need to be difficult. It is only a matter of time before becoming sustainable isn't an option anymore – it is something we must do if we want to protect life on Earth.

What are the benefits of sustainable living?

There are numerous benefits to living sustainably, including a reduced cost of living and living a healthier lifestyle. Not only does sustainability improve the planet, your quality of life can actually improve!

Here are some of the top benefits of living a sustainable lifestyle – for beginners especially.

Simplify your life: Whenever you start living sustainably, you start to question just how much you need and what you

feel comfortable living without. Whenever you use, need, and want less, you simplify your life. This allows you to make time for the things that truly matter to you, such as family, friends, and self-care.

Saving money: Sustainability actually saves you money in the long run and helps you stop mindlessly spending. You become a much more conscious consumer and save money when you rely on yourself for things like food, water, and energy.

Better planet for future generations: Every action you take now helps your children, grandchildren, and all future generations have the resources they need and the quality of life that was afforded to you. Sustainability helps future generations thrive.

Living a healthier lifestyle: When you start living in an eco friendly way, you tend to get rid of the toxins in your home, which helps you live a healthier lifestyle. Taking care of the planet directly impacts and improves your health! You could also lose weight and increase your fitness by forgoing unhealthy foods wrapped in plastic or taking the stairs instead of the elevator. Win-win!

Increased confidence: Sustainability can help improve your confidence when you rely on yourself instead of others for your wants and needs, and can help your self-esteem when you become part of a global movement.

Less waste: When you live a sustainable lifestyle, you end up creating less waste, which helps reduce your carbon footprint. You may buy fewer clothes, waste less food, and generally consume less, if approaching sustainable living for beginners. If you search through #zerowaste or #ethicalfashion on Instagram, you can find some brilliant companies and methods of reducing waste. NB: I hope these benefits help motivate you to start your sustainable living journey, and to keep motivated when things get tough!

What are some examples of sustainable living?

There are many ways that people are living sustainably! Some measures are more extreme than others, but they are doing their part to protect the planet and are still thriving. Here are a few examples of sustainable living. Check them out to see if these would be viable options for you!

Reduce, reuse, recycle, & re-purpose: This method is a great example of sustainable living! When we reduce our use of resources, re-use what we can, recycle what we don't need, and re-purpose old items into new ones, we help improve the environment. The best part about this example is that it is practical and something everyone can do to live more sustainably.

Tiny home living: There are many people around the world that have seen the benefits of living in a smaller home. These people are enjoying a reduced cost of living, are using fewer resources, and are more energy efficient than those in larger homes. Not only that, but they are living simple, happy lifestyles that focus more on community, family, and experiences rather than material things.

Gardening: Starting a garden is another example of sustainable living. The benefits of starting your own garden are enormous, such as knowing exactly where your food is coming from, saving money by cutting out the middleman, and increased confidence in acquiring a new skill. This example is also practical since you don't need a lot of resources or space to get started growing your own food.

Ditching the plastic: Limiting or completely stopping the use of plastic has become more and more popular and for good reason. Plastic can take thousands of years to decompose, and often contains harmful chemicals that are not good for the environment or animal life. Many people have traded in plastic shopping bags for cloth bags, plastic straws for metal straws, and bring their own reusable containers to the store to buy products in bulk.

Driving green: Another example of sustainable living would be trading your car in for something that is green and eco-friendly. Some people also bike or walk to their destination if the commute is close by. Whenever you can, try to use these options, or carpool if possible.

Practical sustainable living tips: how can I be more sustainable?

If you are ready to live a sustainable lifestyle but not quite ready to sell your house or trade in your car, there are many practical, simple steps you could take today that would greatly benefit the planet and your lifestyle. Kick off your sustainable living journey with easy ways to live more sustainably.

Go Vegan: One of the best things you can do for the environment is to go vegan. If you aren't quite there yet, then consider going meatless for a couple of nights during the week. Learner Vegan has some great advice on how to start! Going vegan has many benefits such as using less water, cost savings, and saving the lives of precious animals.

Change the light bulbs: Consider switching to LED light bulbs. These bulbs are energy efficient, meaning you don't have to swap them out as often as conventional lighting. There are so many different kinds of LED lighting options that you are sure to find something that fits your needs and the style of your home.

Don't use plastic bags at the grocery store: Plastic is extremely harmful for the environment, so it should be our top priority to use it less and less, or go completely plastic-free. Plastic bags at the grocery store are clogging up our landfills and endangering animals on land and sea. Whenever you go to the grocery store, bring cloth bags with you to hold groceries instead. Keep them in your car with you at all times so you never forget them!

Hang dry clothes: Save money and promote energy conservation by line-drying your clothes instead of throwing them in the dryer. You will greatly reduce your carbon footprint and protect your clothes from harsh heat, which helps them last longer as well.

Turn off all devices at night: Turn off all devices at night, including computers and wifi as well, in order to save energy! Don't forget to also turn off all the lights whenever you leave a room.

Avoid buying plastic wrapped products: It seems like plastic is a daily facet in our lives and impossible to escape. However, limiting your use of plastic isn't as hard as you think. You could switch from unhealthy plastic wrapped foods to whole foods, swap out plastic straws for metal, or stop buying water bottles and use a stainless steel bottle instead. This helps protect the environment and your wallet!

Shop second-hand for clothes (or shop ethically): Shop at local thrift stores when you need new clothes. Most thrift stores have name-brand clothes in like-new condition. This will help you save money, fossil fuels and water resources that come from shopping at the clothing store. If you want brand new items, then please shop ethical, sustainable clothing brands. These brands make sure that not only are they conscious of the amount of resources used, but that their workers are receiving fair pay and the companies are enforcing ethical labour practices.

Opt to receive digital files and go paperless!: Whenever you are able to, opt to receive digital mail and notices. Most banks and other companies have a green initiative and offer to send you important emails instead of paper notices. Sustainable living for beginners can be as easy as this!

Make your own homemade cleaning products: Most conventional, store-bought cleaners contain harmful chemicals that are not only bad for the environment, but bad for your health as well. Consider making your own green, homemade cleaning products using natural ingredients such as lemons, lavender, vinegar, and baking soda. This will help you save so much money, since store bought cleaners can get very expensive.

Final thoughts

This is just the tip of the iceberg when it comes to sustainable living.

Not only is it important for the future of our planet to live sustainably, there are practical ways to get started and so many benefits for ourselves and future generations as well. So get going using this guide to sustainable living for beginners!

Tell me, are you already doing any of these things in order to live a sustainable lifestyle? If not, is there anything you are willing to try?

October 2019

An ethical future could make life harder for the poorest – but it doesn't have to

An article from *The Conversation*.

THE CONVERSATION

By Donald Hirsch, Professor of Social Policy, Loughborough University

The British supermarket chain Morrisons recently announced that it will only sell free range eggs. This is a telltale example of how business and government are starting to do more to encourage or require ethical consumption in the UK.

The government recently announced that solid coal and wet wood can no longer be used in domestic burners and

fireplaces. And the Chancellor of the Exchequer is reported to be considering an increase in fuel duty in the budget, in line with carbon reduction objectives.

All these things have the potential to increase basic living costs, including for worse-off households who are already struggling to make ends meet. Poorer working-age households have seen their buying power squeezed in recent years. For example, my team's research on minimum household living costs in the UK shows that these typically rose by 8%-12% from 2015 to 2019 (varying by household type). This is the same period over which benefits and tax credits were frozen in cash terms.

Growing use of food banks reflects the vulnerability of households living on the edge when there is nothing to fall back on if things go wrong. For such families, increasing the cost of the basics even by what may seem like small amounts can cause additional hardship.

In the coming years, there will be growing pressures to increase the cost of basic food and fuels in response to a range of environmental and ethical concerns. From reducing energy use through pricing, and switching to less polluting modes of consumption, to responding to ethical concerns around issues such as animal welfare, environmentally responsible farming and fair trade.

Where this involves regulation or supply changes like Morrisons's decision on eggs, rather than free choices by consumers about buying ethically, worse-off households will be vulnerable to further increases in living costs. Food and energy form a disproportionately high proportion of overall spending for the worst-off.

This has the potential to create tensions between the interests of needy groups and a range of ethical concerns. But this does not need to be a straight trade-off if policies can be designed to take both aspects into account. To understand some of the complexities, consider the case of meat.

Red meat

Most people in rich countries eat far more red meat than they need to or is good for the planet. But when we asked members of the public if they would view, for example, eating less meat as compatible with an acceptable living standard, they were resistant. This is partly because some meat, such as chicken and beef, has become relatively cheaper in recent years and so a more economical way of feeding the family.

Our research showed that people are heavily influenced by price. They were most committed to doing things to 'save the planet' when this also saved their wallets, such as reducing their energy use after gas and electricity prices rose.

Over the past 15 years, British lamb prices have risen much faster than other meats, and lamb consumption has fallen. This shows that over the long term, eating patterns can gradually change, influenced by price. This makes our relatively recent shift to a more meat-based diet reversible – a change that would also help improve living standards by making people more healthy.

But this doesn't address the immediate issue of how to avoid creating extra hardship by raising the price of goods in the interests of ethical consumption. In doing so, it is helpful to think about the meaning of terms like 'food poverty' or 'fuel poverty'. To what extent is difficulty affording such items driven, on the one hand, by low income, or on the other by high costs?

The official definition of fuel poverty was changed in 2012 to cover only cases where households have both relatively low income and relatively high fuel requirements. In these cases, an important part of the solution is to get a household's fuel costs down, for example through subsidies to improve the energy-efficiency of people's homes targeted at those in fuel poverty.

Maintaining living standards

In cases where the affordability issue is simply down to having very low income, solutions should be to help improve people's earnings and public benefits. Looked at another way, if the only way people are able to afford to eat is by selling them ultra-cheap food produced in unsustainable ways, the solution is not to lower standards but to raise incomes.

Yet given that there will continue to be people finding it hard to make ends meet, those introducing ethically-driven changes affecting the price of basics need to think actively about what can be done to mitigate the impact.

One option is to consider how, when one product is withdrawn or made more expensive through taxes or regulation, affordable substitutes can be made available.

Following the recent banning of solid coal and wet wood for stoves, on which some households rely for heat, it will be important for government to ensure that alternatives, such as kiln-dried wood, are both available and sold at reasonable prices, even if this may initially involve a degree of subsidy or regulation. Similarly, were regulations on caged hens to be tightened, governments may consider whether additional subsidies could help to limit the impact this has on price.

The media and public have grown wary of excessive intervention in free markets, but could come to accept that more interventions are needed to make ambitious commitments on emissions reduction a reality. In such a future, it is crucial that those who intervene to set ethical and environmental standards are also more active in ensuring that this does not increase hardship for the most needy households.

2 March 2020

'I live a zero-waste life – here are my tips to reduce your impact on the planet'

Kathryn Kellogg says she saves thousands each year through her zero-waste lifestyle.

By Mora Morrison

In brief

♦ Being zero-waste adds 'two minutes – maximum' to Kellogg's day

♦ All of Kathryn's rubbish for a year fits into a regular jam jar

When Kathryn Kellogg was at school, she had a peanut butter and jam sandwich for lunch every day. Each sandwich was wrapped in a plastic bag, and she threw away every single one. She's worked out that she sent around 2,275 sandwich bags to landfill.

Now, Kellogg starts every day with a smoothie (made from cauliflower - there's no accounting for taste). This, along with all of her other purchases and unlike her childhood lunches, doesn't produce any waste.

Kellogg is one of a growing number of people - many of them millennial women - who are part of the zero-waste movement, and her rubbish for the past year fits inside a regular jam jar.

The 28-year-old spends five hours a day on her lifestyle blog, Going Zero Waste, where she writes about, among other things, home-made hairspray, buying in bulk and the sharing economy.

And now she has compiled all she has learnt on this zero-waste journey, which she admits is not perfect, to produce a nifty guide called *101 Ways to Go Zero Waste*.

Just a few small changes

'Living in California near so many other people trying to reduce their waste makes it much easier for me,' she says. 'But I still think that everyone could reduce their trash by about 75 per cent just by making a few small changes.'

Kellogg decided to change her habits after a breast cancer scare when she was 20. She adopted a more natural lifestyle which, she says, 'has had a huge impact in linking my own health and the health of the planet'.

'One of the first things was to change cleaning products,' she says. 'It's so much cheaper and I would much rather wash some lemon juice and vinegar down my drain than something with 17 words that I can't even pronounce.'

Reduce what we need, reuse what we can

Many of her solutions to cut waste were commonplace before the era of plastics and disposables. Think cloth napkins and grocery bags, and glass or stainless steel containers for leftovers.

Kellogg visits farmers' markets so she can buy groceries free from packaging, she carries a reusable coffee cup and shops with cloth bags made from old sheets, which she fills with dry goods such as rice and pasta.

Her aim is to 'reduce what we need, reuse what we can, send as little as possible to be recycled, and compost what's left over'.

Fast facts: recycling

♻ Two-thirds of all UK households are not sure which bin to use for one or more items

♻ Almost half of all UK households throw away one or more items that could be recycled

♻ Just over two-thirds of all UK households put things in the recycling which cannot be recycled

♻ If you scrunch paper and it doesn't spring back, it means you can recycle it

The following items **cannot** be recycled (although you might think they can):
- Sweet packets or wrappers
- Bubble wrap
- Laminated pouches - such as for cat food and coffee
- Plastic toys
- Blister packs for medicines (like paracetamol)
- Toothpaste tubes
- Expanded polystyrene (used in packaging to protect your products)

Many zero-waste pioneers trace their moment of realisation back to Bea Johnson, a charismatic mother of two who lives in Mill Valley, California. Johnson started blogging about her own lifestyle in 2009 and confirmed her status as a role model for the movement in 2013 with the publication of *Zero Waste at Home: the Ultimate Guide to Reducing Your Waste.*

A life of 'being rather than having'

'It's made time in our life for what matters most, a life based on experiences rather than things, on being rather than having,' Johnson said at a TED conference in 2016.

She has also worked with corporations, spoken at the United Nations and advised retailers such as Ikea to replace plastic utensils with durable (and recyclable) metal alternatives.

Each of her family members can fit all of their possessions inside a suitcase and if they go on holiday they simply pack up their worldly goods and then rent out their empty house - using this income to fund their trip.

As well as simplifying her relationship with 'things', Kellogg says she is saving around $5,000 (£3,800) a year and that being zero-waste only adds 'two minutes, maximum' to her day.

The 'Blue Planet' effect

Her quest to reduce landfill comes at a time when the world is swamped in rubbish and plastics. According to government statistics, the UK generates around 222.9 million tonnes of waste each year.

In 2017 the total waste from households in England alone was 22.4 million tonnes, which is the equivalent of 403kg per person - roughly the weight of two motorbikes.

The Blue Planet effect, which has been felt since Sir David Attenborough's nature series first aired, has also helped raise the alarm about the problem of plastics - which, according to the Ellen MacArthur Foundation charity, will overtake the number of fish in the ocean by 2050.

Zero-waste influencers

In response, zero-waste influencers have utilised social media in an attempt to bring their movement into the mainstream. The #ZeroWaste hashtag has more than 2.6 million Instagram posts, with prominent millennials such as Andrea Sanders of Be Zero Waste Girl and Lauren Singer of Trash is for Tossers leading the way.

They embrace a sleek and modern aesthetic over the tree-hugging stereotypes that are sometimes associated with the movement.

Kellogg is enthusiastic about the movement's potential, but she has also weathered her fair share of criticism along the way.

'There are people who think I am too extreme and those who view me as not extreme enough,' she says. 'Some people can't believe I'd take a flight home to see my mum [in Arkansas] or someone might say: 'You wrote a book with actual pages, you're killing trees'.'

But she is happy with her choices.

'This isn't a crash diet - and something you can burn out on. It's all about balance. No one will leave this planet without having an impact on it but you get to decide how much of an impact you're going to have and, for me, I'm always trying my best to have the lowest impact possible.'

6 September 2019

Which supermarkets use the most and least plastic?

Find out how your favourite supermarket fares when it comes to problematic plastic packaging.

By Angela Terry

Plastic has never been as unpopular as it is right now. For decades it was lauded as a miracle material – cheap, versatile, convenient – and the world gobbled it up tonnes and tonnes at a time. Now, though, we have a problem.

The plastic we use today will exist on the earth long after we're gone. Indeed, the plastic we used as children will still be floating around somewhere on the planet. Many types of plastic are difficult – if not impossible – to recycle, so the material is doomed to clog up landfill, spoil natural environments and, as devastatingly depicted on the BBC's Blue Planet series earlier this year, irrevocably damage wildlife and ecosystems.

Then, of course, there's all the energy and natural resources used to create the material in the first place – and it's a lot. Making enough plastic water bottles for the United States, for example, requires 17 million barrels of oil and enough energy to power 190,000 homes for a year. And that's just one type of plastic product in one country! To fix climate change we have to keep fossil fuels in the ground.

There are lots of ways we can help tackle the problem, from buying refillable water bottles to using natural materials, but one of the biggest ways to make an impact is by changing our shopping habits. According to Greenpeace, UK supermarkets are responsible for churning out at least 59 billion pieces of plastic every year, so choosing stores and retailers that are themselves committed to making a change is a great step to take.

And it's something they're all getting on board with – however slowly – as research shows that reducing packaging waste will be shoppers' biggest concern in the coming years. But even though supermarkets are now turning their attention to reducing single use plastic and getting rid of non-recyclable plastic entirely (and of course, shouting about their good efforts in the process) the question is, which supermarkets are really making a difference?

As research from Greenpeace shows, there's a pretty significant variation when it comes supermarkets' plastic-reduction efforts. Its plastic league table – which covers the UK's top 10 supermarkets – takes a number of factors into account, including measures to influence suppliers and be transparent with customers, as well as getting rid of plastic itself. Here are the fast facts.

The best supermarket for plastic reduction activity is…

Iceland. The family-friendly frozen food chain is taking bigger strides than most in reducing single-use plastics and

eliminating non-recyclable plastic packaging. It's also doing a fair job of influencing its suppliers to do the same, and being clear and honest with its customers about what it is doing. However, while Iceland topped the charts, it still only achieved a score of 5.7 out of 10 – meaning there's plenty more to be done.

The supermarket doing the least to combat plastic is…

Sainsbury's. With a low score of 3.2 out of 10, Greenpeace says that this well-established retailer is at the bottom of the pile when it comes to policies designed to reduce plastic waste. On the upside, it is one of just four supermarkets that have

How did other supermarkets compare?

With a top score of 5.7 and a bottom score of 3.2, there's not a huge amount of difference among supermarkets, but some have performed better than others. Here's how they stack up:

Iceland: 5.7/10

Morrisons: 5.3/10

Waitrose: 4.7/10

M&S: 4.6/10

Tesco: 4.5/10

Asda: 4.3/10

Co-op: 4.2/10

Aldi: 4.1/10

Lidl: 4.1/10

Sainsbury's: 3.2/10

initiatives underway for refillable and reusable packaging – the other three are Morrisons, Waitrose and Tesco. Iceland, by contrast has no current plans for such a scheme, nor does Co-op.

When will supermarkets be plastic-free?

Plastic helps to keep some food fresh, makes goods easy to transport and it's cheap, but anyone trying to go plastic-free knows there's far too much of it in supermarkets. Plus, many supermarkets still charge a premium for loose fruit and veg! But some supermarkets have ambitious targets that mean we'll see meaningful change within the next decade.

A number of UK supermarkets have signed the UK Plastics Pact. This is a voluntary scheme where companies and retailers commit to reducing the total amount of plastic packaging in their business models, and to make unavoidable plastics recyclable.

Other supermarkets have set their own targets. Iceland, for example, wants to reduce its own-brand single-use packaging by 100% come 2023. Aldi, Sainsbury's and Tesco want to reduce all of their plastic packaging by 50% by 2025 at the latest. Waitrose, meanwhile, has trialed a plastic-free supermarket in Oxford, which proved very popular with customers. Others are making a less ambitious approach – Lidl is aiming to reduce its own-brand single-use packaging by 20% by 2022.

However, while some supermarkets are moving slower than others – and none of them are moving particularly quickly – they are at least moving. Without consistent pressure from their customers, their scores could have been even lower – so keep saying no to plastic, and one day supermarkets will stop giving us the option altogether.

1 August 2019

www.onehome.org.uk

Plastics killing up to a million people a year, warns Sir David Attenborough

By Anne Gulland, global health security correspondent

Sir David Attenborough has warned that the growing tide of plastic pollution is killing up to a million people as year as well as having devastating consequences on the environment.

A report on the impact of plastic pollution, one of the first to document the impact of discarded plastic on the health of the poorest people in the world, estimates that between 400,000 and one million people die every year because of diseases and accidents linked to poorly managed waste in developing countries.

Sir David, whose Blue Planet TV programme alerted the world to the damage plastic was wreaking on the oceans, says that the effects of plastic pollution is an 'unfolding catastrophe that has been overlooked for too long'.

He said it was time to act 'not only for the health of our planet, but for the wellbeing of people around the world'.

'We need leadership from those who are responsible for introducing plastic to countries where it cannot be adequately managed, and we need international action to support the communities and governments most acutely affected by this crisis,' he said.

The report, by charities Tearfund, Fauna & Flora International and WasteAid, warns of a growing public health emergency, affecting the poorest and most vulnerable people in the world.

Just one in four people around the world have their rubbish collected so plastic and other waste often ends up discarded in the environment, blocking waterways and drains. This leads to flooding, which, in countries with poor sanitation, leads to outbreaks of cholera and other diarrhoeal diseases, as well as drowning.

Several severe floods have been attributed to blocked waterways, including a flood in Accra, Ghana in 2011 where a resulting cholera outbreak killed 100 people.

Discarded plastic also provides a fertile ground for disease vectors such as malaria- and dengue-carrying mosquitoes which breed in the rainwater collecting in waste.

The report also highlights the link between plastic waste and air pollution. For many in low and middle income countries the only way to get rid of plastic and other waste is to burn it, releasing toxic fumes into the air.

The World Health Organization estimates that air pollution is responsible for 3.7 million premature deaths a year, and recent estimates suggest that open burning could be responsible for as much as a fifth of this death toll.

The report also highlights the unknown impact of microplastics – small pieces of plastic less than 5mm in diameter – on human health. The presence of microplastics in the oceans has been well documented but the impact of microplastics in soils, sediments and freshwater is less well recognised. There are fears over the impact on human health if microplastics are ingested, although there is little hard evidence in this area yet.

The report says large multinationals such as Coca Cola, PepsiCo, Nestlé and Unilever need to do more to both reduce the amount of single-use plastic that goes into their packaging and help low and middle income countries with their waste management.

It says that companies should report on the number of single-use plastic products they use and sell in each country around the world by 2020. And, by 2025, they should reduce this amount by half and use reusable containers instead.

Joanne Green, senior policy adviser at Tearfund, said that the big multinationals should not be selling goods packaged in single-use plastics in countries that did not have the waste management systems to deal with them.

'People in developing countries have no choice but to burn or dump plastics. It's not a responsible business model and we're asking customers to sign a petition asking companies to change.

'We want companies to clean up their act and stop producing single-use plastic and collect one piece of plastic for every one they produce,' she said.

She added: 'We are also calling on donors to increase aid to government so they can support countries to implement things like bans. But this is a longer-term solution to an urgent problem. Plastic waste is increasing and the generation of plastic is increasing rapidly.'

14 May 2019

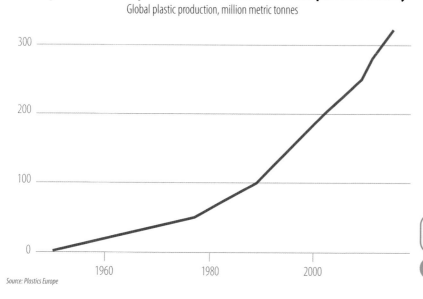

More plastic has been produced this decade than in the previous century

Global plastic production, million metric tonnes

Source: Plastics Europe

What is the problem with the Blue Planet effect?

Has the Blue Planet effect really made a great change to our world, or have we shifted so much focus onto plastic that we've neglected the wider picture of waste management?

We're waging war on single-use plastics, such as straws and bottles, after witnessing marine life struggle and suffer through swathes of plastic. We're ditching plastic bags after seeing footage of plumes of the stuff floating through the sea like stranger forms of jellyfish. After millions tuned in to the highly successful *The Blue Planet II* in 2017, our sense of responsibility for plastic waste has never been greater.

Or, at least, that's what various news outlets seem to be reporting.

With the likes of Aldi and Lidl making moves to improve their packaging for the sake of the environment, their brand image, and at the end of the day, their sales, it seems like everyone is rallying together. The government has even announced an earmarked £61.4 million for funding the battle against plastic.

Helistrat, a UK-based management consultancy that delivers sustainable resource strategies, have observed a notable increase in the number of clients approaching them with a focus on what can be done to deal with plastic waste. The challenge, they say, is the need to 'design out' plastic clashes with health and safety restrictions revolving around on-the-go products.

Between the highlighting of the plastic problem, and the recent ban on plastic waste imports to China, it's never been more crucial for the UK and the world to create a far stronger recycling system and capability. Closed-loop solutions are proving popular, with retailers leaning more towards compostable and biodegradable items like bags and coffee cups in order to phase out plastic.

These are all positive steps, but it's not enough to simply replace plastic. The oceans will not be cleaner, the landfills will not be reduced, by simply banning plastic — we need to build the appropriate infrastructure to process these alternatives properly.

We also must avoid the trap of forgetting the impact of other waste material on our planet's health too. It's easy to get carried away in the panic of plastic caused by the media, but plastic is not the only culprit causing problems for our environment.

This is the problem with the 'Blue Planet effect' — we're pouring so much care and concern into plastic, labelling it the ultimate enemy of the environment, that we're not seeing the bigger picture. Yes, plastic is a huge contributor to the deteriorating natural world, but it is one symptom of a greater problem.

For example, one key demand for change as a result of the 'Blue Planet effect' was the use of single-use plastic straws. As a result, consumers demanded a change in the use of plastic straws, spurred by the horrific figures of a single straw taking so many hundreds of years to break down. Plastic straws were, for a while, public enemy number one.

But according to United Nation figures, of the 9 million tonnes of plastic that ends up in the ocean every year, around 2,000 tonnes of that is made up of plastic straws. That's just 4% of the total plastic waste, and there's the other 8,998,000 tonnes of plastic waste left to consider.

Yes, the change in attitude towards single-use plastic straws is certainly a positive step, and definitely beneficial in removing this utterly unnecessary product, but we must be vigilant and not stand stagnant in an obsession with plastic straws. With one item duly being addressed and the awareness raised of its impact, let's now give focus to other elements that need work, and give them the same spotlight we did with plastic straws.

What about workplace recycling? Does your workplace have segregated bins, or perhaps you're in an industry that makes frequent use of larger waste removal via skip hire? Does your workplace have full transparency from their waste management company as to where that waste goes after it's moved off-site?

These issues might not seem as dramatic and interesting as the shocking images of sea life injured by straws, but their end result impacts the world all the same.

The fact of the matter is, the world needs a better way to deal with its waste problem, and though plastic is certainly a key player, it's not the only villain at play. We mustn't rest on our laurels now that the problem of plastic has been highlighted and changes are being made for it — this is just the first step. It is a big step, and it's in the right direction, but the journey is as important as the embarking.

12 August 2019

Is your reusable tote worse for the environment than a plastic bag?

By Rosie Frost

Plastic bags have become a bit of a taboo subject in recent years. Bans have been brought in across the globe to try to limit how many of these environmentally disastrous pieces of junk end up littering our world for virtually forever. Even some of the most prolific waste producers are making plastic bags a thing of the past with China looking to ban non-biodegradable bags by the end of 2020.

Despite our new-found distaste for the single-use carrier, the Earth Policy Institute estimates that nearly a trillion plastic bags are still used worldwide each year. That works out at 2 million every minute. As they will never biodegrade, bags often make their way out of our rubbish and into the world, end up caught on trees or floating in the ocean to poison, choke and entangle animals. Even when they do break down, the tiny pieces, called microplastics, have been found in our food and drinking water supplies.

As we have become wise to the impact of this waste, there has been a shift toward more eco-friendly options to stop our weekly shop from contributing to swirling islands of plastic growing in our oceans. As supermarkets look to adopt all kinds of seemingly sustainable swaps from paper to reusable alternatives, research has shown that some of these options might not be so green after all.

Pitting plastic against paper

A Danish study from 2018 took a look at the environmental impact of different carrier bags throughout their lives, from harvesting the raw material to you putting your carrots and cauliflower in them at the supermarket checkout. Pitting traditional single-use plastics against alternatives like more heavy-duty 'bags for life', paper and cotton, the team used 15 categories of environmental impact to work out how many times a bag needed to be reused.

The study created a lot of controversial headlines for the claim that cotton tote needed to be used thousands of times to have the same impact on the environment as a single-use plastic bag. If you look a bit closer, however, it doesn't really spell the end for this eco-warrior favourite. The reason for these very large figures wasn't water usage or CO2 emissions but instead damage to the ozone layer; damage that was caused by gases used to help transport the fossil fuels powering the pumps that water the cotton plants.

It is an unfair generalisation, though, as not all cotton is produced in the way they describe. Any impact on the ozone layer could be easily mitigated by switching irrigation away from fossil fuels to more renewable sources. The Danish

Sales of 'bags for life' increased by 26% in the UK in 2019

The total number was enough for 54 bags per household, per year

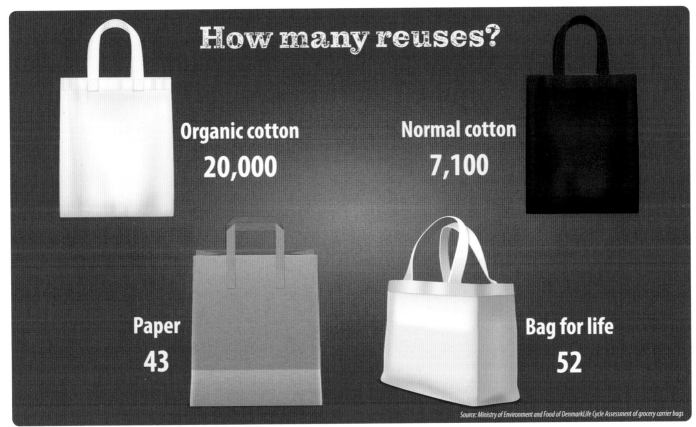

How many reuses?

Organic cotton
20,000

Normal cotton
7,100

Paper
43

Bag for life
52

Source: Ministry of Environment and Food of Denmark Life Cycle Assessment of grocery carrier bags

study reveals that if climate change alone is considered, the minimum number of re-uses goes down to 149. That's just two trips to the shop a week for one and a half years, a far more achievable number.

A bigger problem than cotton could be our growing addiction to the reusable 'bag for life'. The Environmental Investigation Agency (EIA) and Greenpeace found that the introduction of a charge in the UK in 2015 might actually have increased the amount of plastic we use thanks to these thicker and heavier bags. Despite an 83% reduction in the use of conventional carrier bags, people are still buying these heavy-duty replacements at a similar rate with 1.5 billion sold in the UK last year.

'No recycling system in the world could deal with the volumes of plastics currently being used in the world every day,' says EIA Ocean Campaigner, Juliet Phillips, 'we need to address the problem and significantly clamp down on single-use plastics.' She explains that the 5-10p charge introduced in the UK was too low to stop customers from treating reusable bags as a throwaway item. 'UK households are getting through 54 'bags for life' each year suggesting that they are being used as a bag for a week,' Phillips says.

Marine pollution matters

There's something major missing from the Danish study as well, the disastrous impact of plastic bags on marine life. So uncontrollable is their spread that in 2019, an American explorer breaking a world record for diving to the deepest part of our oceans, the Mariana Trench, found a plastic bag at a depth of nearly 11km. That means that plastic waste has made it to parts of our planet that humankind has not even been to yet.

Although perhaps not as damaging to the environment in the way the Danish scientists measured, conventional plastic

bags easily find their way into the sea where they can cause the death of marine animals. Several species of sea turtles are particularly fond of feasting on jellyfish and a floating plastic bag bears a striking and dangerous resemblance to their favourite food. A recent Greenpeace study in French Guiana stated that an estimated 50% of all sea turtles have ingested pieces of plastic, something that makes them far more likely to wash up dead on our beaches.

It isn't just whole bags drifting with the ocean currents that cause problems, however. These single-use items may never biodegrade but when they eventually break down they become microplastics. At least 50,000 particles of microplastics make their way into our bodies each year says a study in the journal Environmental Science and Technology. Chemicals like flame retardants and antimicrobials can leech out of the tiny bits of plastic wreaking havoc on our endocrine systems.

'We need system change, not material change,' says Phillips, 'all materials have an environmental footprint and switching to paper bags is a false solution to the pollution crisis.' It turns out that, instead of swapping to an apparently eco-friendly alternative that just shifts the problem elsewhere, it's better to just stick with reusing the bags you've already got as many times as possible.

22 January 2020

The importance of sustainable fashion in a sustainable world

By Emma Foster

If we are to see any major changes towards a sustainable future, our cultural attitudes surrounding the fashion that we consume and dispose of needs to change.

Today, there is a growing awareness about the lifestyle factors that are linked to pollution and climate change. These include the consumption of animal products and the usage of plastics. However, there remains another factor that needs attention that is vital in ensuring a truly sustainable future: the fashion industry and our ferocious appetite for constant clothing consumption.

The environmental costs of clothing production are immense. Devastating amounts of water are needed to grow cotton, the pesticides used to speed up the rate of growth are harmful to the land and to human health and factories belch out toxic chemicals from clothing dyes. The global activism movement focusing on sustainable fashion, Fashion Revolution, has reported that the fashion industry is responsible for 20 per cent of global water pollution. This makes rivers, lakes and seas in countries of production extremely harmful to the surrounding inhabitants and ecosystems.

Furthermore, many of the world's clothes are made by an overwhelmingly female workforce, who are forced to work in dangerous, life-threatening conditions. Fast fashion companies continue to exploit these employees, paying them as little as possible in order to keep the most profit when the clothing is sold.

In recent years, there has been a growing trend towards buying second-hand clothing, rather than always turning to the high street. Big names in the fashion industry have also been raising the issue of sustainable and ethical clothing, such as Stella McCartney and Oxfam who held a second-hand show at the most recent London Fashion Week.

Despite this growing public interest in sustainable fashion, the continuous and regular consumption of new clothes remains the Western cultural norm.

Fashion brands are able to accelerate their harmful production as they are catering to a demand for the newest, trendiest products at the lowest price. Fashion advertising, celebrity culture and social media all feed into this materialist ideal that encourages consumers to show their social status through the clothes they wear, with no regard to where these items may have come from or how they were made. The extraordinarily cheap price and disposable quality of these fast fashion garments make it all too easy for consumers to wear them a couple of times, then replace them with next season's trend after a few weeks without batting an eyelid.

In a panel discussion at the 2014 Copenhagen Fashion Summit, Livia Firth summed up the downfalls of this cultural phenomenon on the consumer end: 'Is it really democratic to buy a t-shirt for $5, a pair of jeans for $20? Or are they [fast fashion brands] taking us for a ride? Because they are making us believe that we are rich and wealthy because we can buy a lot, but in fact, they're making us poorer, and the only person who is becoming richer is the owner of the fast fashion brand.'

Textile disposal is another major environmental concern in the lifecycle of these garments. This is a problem that is widely hidden and therefore unconsidered by the consumer when purchasing new clothing. Items that are thrown away or are returned to an online retailer often end up in landfill, releasing toxic chemicals during the long decomposing process. Even more shockingly, up to 90 per cent of charity shop donations are sent to developing countries where most items go unwanted and, due to their cheap price, create unbeatable competition for the businesses of local garment makers.

Although there is no simple solution to the damage caused by the fashion industry, changes in consumer habits have the power to make a difference. By switching the demand towards second hand or sustainably and ethically made clothing, pressure can be applied to fast fashion brands to improve their production methods. Online marketplaces now exist for buying and selling good quality second-hand clothing, such as Depop and Edinburgh based start-up One Cherry. Furthermore, by buying from charity shops as the first port of call, the amount of textile waste being sent from these shops to the landfill or to developing countries can be reduced. By choosing good quality sustainably made clothing that is meant to last, or by mending and upcycling the clothes that we already own, we can prolong the lifespan of clothing, preventing it from becoming waste.

Throughout history, our cultural attitudes towards fashion govern the way in which our clothes and accessories are made, sold and disposed of. Consumers must recognise fashion decisions, not as an inconsequential part of our modern lifestyles that changes with every season, but as an environmental and human responsibility that could greatly affect the future of our planet.

29 April 2019

Why you should care about sustainable fashion

By Emily McCoy

Only a tiny percentage of the money we pay for clothes ends up in the hands of those who made them. The fashion industry is notorious for exploiting it's garment workers and cotton farmers who are often left invisible at the end of a long and complex supply chain.

Why should we care about sustainable fashion? Take it from someone who knows best what sustainable sourcing can do. Phulme is a cotton farmer working for a Fairtrade certified co-operative.

Phulme lives in the remote Bolangir district in the Odisha state of eastern India where her community, the Pratima Organic Growers Group, farm sustainable cotton. Her co-operative is made up of over two thousand individual members spread across 72 villages. Until now, they received little contact from the outside world and lived in relative poverty.

The Pratima co-op have been Fairtrade certified since 2010. They are a democratic society and every three years choose a chairperson to speak on behalf of, and lead, the group. That person is Phulme Majhi, and she talks to us about being a woman leading the Pratima Group, and explains some of the challenges, both environmental and social, that they face.

Sustainable fashion empowers women

'I was elected as the chairperson of our co-op. There was a voting process between a male farmer and myself, and I won! I was very apprehensive in the beginning, being a woman, I wondered how I would manage to deal with the men on the board. I was very afraid and scared. But I am not afraid anymore.

'Now we can walk shoulder to shoulder with men. We [women] have access to finance and the confidence to handle our own finances, whereas in the past we relied on men. We get training and we can meet visitors – this gives us more confidence in ourselves. We are able to fulfil our needs ourselves, and not be dependent on men.'

Fairtrade provides co-ops like Pratima with training and support to give farmers the opportunity to improve their lives. Phulme was also involved in a women's group with other women in her village.

'We were educated about the benefits of a women's group and asked to set up our own. I spoke to the other women and explained the benefit we would get from this group. This group taught us to realise our own strength, and that we can go outside of our homes and village, that we can go to the bank and attend meetings.

'Before I was restricted to my house and I did not do anything. I used to do domestic work like cooking and looking after the children, and helping at the farm – that was all.'

Fairtrade helps farmers and workers learn about women's empowerment, making sure people have basic human rights like education and equal rights. With money from their Fairtrade Premium, Pratima also gives cash scholarships to 600-700 school students each year. Phulme speaks enthusiastically about how the women's group she is part of gave her the confidence to become the chairperson for the whole society.

Sustainable fashion supports farmers

In Odisha, cotton farmers face drastic environmental challenges. It is a hilly area, where only cotton can grow on the steep slopes and poor soils. They grow a little rice and some pulses as staple crops in flatter places, but rely on cotton for their basic needs. Sometimes there is no opportunity for any crop during the second part of the year and people may have to leave their homes and look elsewhere for work.

'There used to be no rainfall. We have a problem with rainfall. We have received facilities to help this – Fairtrade has helped to build a water storage unit to preserve rainwater.

'We have also learnt about agricultural practices like composting, which we were not doing ten years ago. We now know about better farming practices, we used to till the land manually and now we have access to tractors. In summers we still face a lot of problems with lack of water. During the rainy season it is OK, but afterwards we are facing a lot of problems.'

Learning and growing

Cotton harvesting takes place over two months, from December to January. Each farm sees two to three pickings and the produce is stored temporarily until the final picking is complete.

The co-op representatives take samples from each farmer for quality checking and then farmers take the cotton to the resource centre, where the collected cotton is taken to the gin for processing. The gin is a machine that separates cotton fibres from their seeds. The resource centre has a weighing machine where the farmers weigh the cotton themselves and then it is again weighed at the gin in front of a farmer representative. Payment is made directly to the farmer's bank account – something Phulme is proud to be able to manage for herself.

During processing, meticulous care is taken to ensure the integrity and traceability of the Fairtrade cotton.

'Fairtrade gives us training in how to tackle our problems. We have also had the opportunity to meet with other groups and businesses. Through this exposure, and being associated with Fairtrade, we have the confidence to speak to people and other officials. Previously we would just depend on what we were told, but now we can decide for ourselves.

'We are a very small village, no-one used to come to meet us, and now lots of people come to see us. We all really like it when someone comes to visit!'

Support cotton farmers like Phulme by buying clothes and homeware made with Fairtrade cotton.

24 April 2019

Is the dress green or red? Planet-friendly couture won't be for everyone but it can lead the way

An article from *The Conversation*.

THE CONVERSATION

By Mark Liu, Chancellors Postdoctoral Research Fellow, Fashion and Textiles Designer, University of Technology Sydney

Hollywood legend Jane Fonda hit the 92nd Academy Awards ceremony stage this week in a beaded red dress by Elie Saab – a gown she had previously worn at Cannes in 2014.

Rising star Kaitlyn Dever walked the Oscars red carpet in a deep scarlet Louis Vuitton dress she told reporters was 'completely sustainable' thanks to fibre technology.

Two very different fashion approaches towards saving the planet – but how effective are they at mitigating the environmental impact of fashion? The first comes from an activist trying to be more sustainable, the second from a designer label making production changes.

Size of the problem

The fashion industry creates in excess of 80 billion pieces of clothing a year and is responsible for 10% of global carbon emissions.

Chemical dyes in fashion create 20% of global waste water and crops such as cotton use 24% of global insecticides.

Microfibres and micro plastics from laundered garments have contaminated our beaches, bottled drinking water, and aquatic food chain.

Fast fashion promotes the reckless over-production and over-consumption of cheaply made clothing, and is reliant on exploiting inexpensive labour.

Less than 1% of fabrics can be recycled and fibre-to-fibre recycling technology is still in its infancy, without the infrastructure to address the vast amount of garments produced.

A 'completely sustainable' dress?

Dever's dress was made from a material called 'Tencel Luxe', with lyocell fibre created from sustainably-grown trees. The wood is cut into tiny pieces, dissolved in a solvent and extruded into a soft cellulose fibre.

This process was first developed in 1972 to create a cheap cotton substitute, often blended into cotton and polyester in inexpensive fabrics. Over time, lyocell improved to make a fabric more appropriate for luxury products.

The sustainability of the fabric has also improved significantly. Today, 99% of the solvent used in the manufacturing process is recyclable. The fabric itself has the same challenges as cotton when it comes to recycling. The length of the fibres breaks down and shortens over time, lowering the quality when recycled into new fabrics – unless virgin material is added.

Indeed, lyocell fabric is far from perfect, and requires a large amount of energy in its production. Just substituting one material for another does not solve sustainability problems.

Dever's dress was made over 1,900 hours by a team of artisans with 14,400 Swarovski crystals and glass beads. A gown this opulent uses so much material, energy and labour that its carbon footprint becomes excessive.

The sustainability measure of a garment must include how it is recycled after its life. At present, some cutting edge fibre-to-fibre recycling technologies exist but Louis Vuitton is yet to offer recycling services.

The Louis Vuitton group hosts the Viva Tech Conference for exhibitors working on sustainable concepts to showcase their developments. It is encouraging that the company's chairman and chief executive Bernard Arnault believes 'sustainable, globalised growth is possible' and a priority for the company.

A second outing

With her commitment to climate change activism, Fonda has said she will no longer buy new clothing.

In a world that demands novelty, Fonda's bold act of choosing an old gown that tastefully fits into today's trends truly brings 'vintage' clothing to the red carpet.

Fonda used her celebrity influence to turn a dress on the red carpet into a political symbol.

Her vow not to buy new clothing was inspired by teenage activist Greta Thunberg, who Fonda said showed that we can't just 'go about our business' in the face of a climate emergency.

No stranger to being arrested, Fonda's real world activism is the type needed to change government policies in ways that rein in the fashion industry.

Activists can petition governments, watch over corporations and form grassroots community groups to organise change. Celebrities who straddle the red carpet and the picket line such as Jane Fonda (or Emma Thompson for Extinction Rebellion) are key to a sustainable fashion industry.

Think before you shop

It is inspiring to see these two stars' dresses become talking points.

It shows the public's growing awareness of climate change and their willingness to change their behaviour to make a

difference. The fashion industry will determine significant aspects of the future of our environment and the lives of over 40 million workers around the world.

From developing new green technology to changing consumer consumption behaviour and outlawing exploitative labour practices that make fast fashion possible, we are still a long way from a 'completely sustainable' industry.

Corporations will need to evolve and adapt to customers who demand sustainability. They will have to offer services that recycle garments after they have been used, and embrace recycled materials. Celebrities bring these issues to the public and give them steps they can take right now.

When asked by a reporter about what people could do right now to be more sustainable, Dever said, 'it's just a matter of letting it be a part of your lifestyle.'

With fashion, I think you can think a little before you buy something brand new, and I think can also support vintage – I think that's really, really important. And also look into the brands you're supporting.

This is a great starting point which hopefully continues into a deeper conversation about how far we yet have to go.

12 February 2020

The story of a £4 Boohoo dress: cheap clothes at a high cost

Online retailers such as Boohoo and Missguided are booming, but critics say there is a hidden price.

By Sandra Laville

It comes in red, mustard and black, in sizes 6 up to 16; the bandeau bodycon minidress, is, according to the online retailer Boohoo, 'perfect for transitioning from day to play'.

It is not so much the styling and colour, but the price of the £5 dress – reduced this week to just £4 – which attracts thousands of the thriving retailer's 5 million UK customers to add it to their online shopping bag, click and pay.

Products and prices like these have driven Boohoo's profits to a record £59.9 million, bucking the trend of struggling high street fashion stores across the country.

Made in the UK, at factories in Leicester and Manchester, the £5 dress epitomises a fast fashion industry that pumps hundreds of new collections on to the market in short time at pocket money prices, with social media celebrity endorsement to boost high consumer demand. On average, such dresses and other products are discarded by consumers after five weeks.

Missguided, an online rival to Boohoo, which also sources products from Leicester, took the low pricing even further this week by promoting a £1 bikini, which proved so popular with customers that the website crashed.

But behind the price tag there is an environmental and social cost not contained on the label of such products. 'The hidden price tag is the cost people in the supply chain and the environment itself pays,' said Sass Brown, a lecturer at the Manchester Fashion Institute. 'The price is just too good to be true.'

A report by MPs into the fashion industry put it bluntly: in terms of environmental degradation, the textile industry creates 1.2 billion tonnes of CO2 a year, more than international aviation and shipping combined, consumes lake-sized volumes of water, and creates chemical and plastic pollution – as much as 35% of microplastics found in the ocean come from synthetic clothing.

Socially, the booming fast fashion industry is often built on low wages paid to women working in factories abroad but also increasingly in the UK, in cities such as Manchester, Birmingham, London and Leicester. There is no evidence that factories in the Boohoo and Missguided supply chains pay illegal wages. However, some garment workers in Leicester are paid an average of £3 an hour – way under the national minimum wage – MPs heard in evidence.

MPs found that the Modern Slavery Act was not robust enough to stop wage exploitation at UK clothing factories. There was a lack of inspection or enforcement, allowing factories – none of which are unionised – to get away with paying illegal wages.

In the UK, we buy more clothes per person than any other country in Europe – five times what we bought in the 1980s, which creates 1.3 million tonnes of waste each year, some 350,000 tonnes of which is dumped in landfill or incinerated.

Yet despite the overwhelming evidence gathered by the environmental audit committee (EAC), ministers this week rejected every recommendation for tackling abuses across the fashion industry, including a ban on incinerating or landfilling stock that can be recycled and a 1p charge on each garment

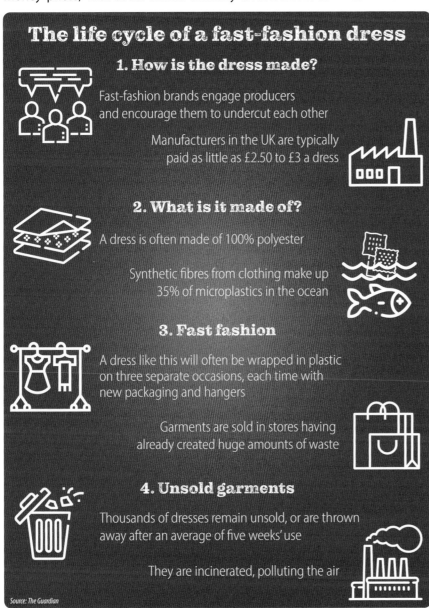

The life cycle of a fast-fashion dress

1. How is the dress made?

Fast-fashion brands engage producers and encourage them to undercut each other

Manufacturers in the UK are typically paid as little as £2.50 to £3 a dress

2. What is it made of?

A dress is often made of 100% polyester

Synthetic fibres from clothing make up 35% of microplastics in the ocean

3. Fast fashion

A dress like this will often be wrapped in plastic on three separate occasions, each time with new packaging and hangers

Garments are sold in stores having already created huge amounts of waste

4. Unsold garments

Thousands of dresses remain unsold, or are thrown away after an average of five weeks' use

They are incinerated, polluting the air

Source: The Guardian

to raise £35 million a year for better clothing collection and sorting, a move supported by many in the industry.

Theresa May said in parliament the government already had measures in place, including a multimillion-pound grant scheme to help boost the recycling of textiles and plastic packaging.

But the response of the government was met with anger by many within the industry, where ethical fashion firms come up against others who produce at lower costs on the back of exploitative wage structures and environmentally damaging production models.

Phoebe English, the English fashion designer, said: 'It is extremely alarming to see that our government has failed to grasp the true extent of the consequences of the fashion industry carrying on the way it is.

'It is a vastly damaging industry that has been spiralling unchecked for far too long. The Earth and the people on it are exploited and damaged at every single step of the chain, and this culminates with unimaginable mountains of unused excess stock or badly made broken waste clothing with nowhere to go other than landfill or incineration.'

English said it was up to the industry to pull together to regulate itself and urgently put an end to bad practice.

But others are less sure that voluntary measures will tackle what they say is a systemic power imbalance between brands and manufacturers, which leads to worker exploitation, or address the enormous environmental footprint of their trade.

'The knowledge is out there. We know the scale of those impacts and we know what some of the solutions are,' said Dr Mark Sumner, a lecturer in retail and fashion at the University of Leeds. 'What we need is the government to come up with some clear policy, whether it's around legislation, or voluntary initiatives to incentivise a change of behaviour by brands.'

With growing scrutiny of their business model, brands are beginning to promote an ethical face. This week Boohoo promoted its recycled brand For the Future, using fabrics made from synthetic waste saved from landfill.

'They are moving in the right direction, but this is a micro-percentage of their output,' said Brown.

While Boohoo is promoting its 'green' credentials, repeated attempts by the trade union Usdaw to talk to workers at its delivery centre in the UK about union rights are being rebuffed by the company.

Mary Creagh, the chair of the EAC, said: 'If brands want to be ethical, they should start by allowing a trade union into their premises rather than barring the door.'

Prof Nikolaus Hammer, associate professor in work and employment at Leicester University, carried out research into the fashion industry in Leicester, where he found employees working in appalling conditions, with no employment contracts, earning on average £3 an hour. It was the brands, he said, who held all the power over manufacturers to keep cutting their prices. 'They will go to the manufacturer and say that person down the road can do it for 1p less, and they get the job,' he said.

Ways to take the environmental heat out of your fashion habit

- Commit to wearing every piece 30 times. If we doubled the amount of time we kept clothes for, we would cut our fashion emissions by 44%.

- Get smart about fibres. Cheap cotton and synthetics come with huge environmental footprints. Cotton uses unsustainable amounts of water and pesticide. Go for hemp blended with organic cotton and silk and lyocel/modal.

- Treat cotton as a luxury fibre. Buy products certified as organic to be free of the pesticide burden and plan to keep them for years.

- Wash clothes less often. The average laundry cycle releases hundreds of thousands of tiny fragments of plastic from synthetic fibres into waterways. Put jeans in the freezer and remove dirt when frozen. Fleeces have been shown to release the most plastic fibres.

- Use a rental service. Peer-to-peer clothes and accessories rental services aim to make use of some of the estimated £30 billion worth of clothing left hanging in wardrobes. Other models include memberships that gives you access to a number of pieces per month.

- Delete shopping apps from your phone and swear off insta-shopping for fashion. Go shopping for clothes in a shop. Feel the fibre, interrogate the structure – especially the seams – and make sure it fits.

- If dry cleaning is a must, use an eco-friendly process: conventional dry cleaning harms the soil, air and water. Use Johnson's GreenEarth system or a dedicated eco cleaner such as Blanc.

He doubted that any greening measures by brands could make them ethical and sustainable while still fulfilling the model of fast fashion.

'The fundamental problem with much of fast fashion is that its social and environmental costs are not taken into account. The environmental costs of materials and fabric are mostly offshored. The production that takes place in the UK often only pays half the legal minimum wage.

'So who is accountable for water and chemical pollution abroad, precarious work and employment, the undercutting of compliant manufacturers, the pollution from delivery and lack of recycling at home? Brands might find it difficult to make £5 fast fashion dresses and be socially and environmentally sustainable.'

Carol Kane, the co-founder of Boohoo, told MPs the £5 dress was a 'loss leader' drawing people to the website. She said the company employed three people in Manchester, three in Leicester, and 10 in China to carry out audits at their producer factories each month to ensure proper working practices.

22 June 2019

UK households waste 4.5m tonnes of food each year, figures show

However, overall levels have reduced by 480,000 tonnes in three years.

By Sarah Young

UK households are wasting 4.5m tonnes of food that could be eaten ever year, new figures show.

The Waste and Resources Action Programme (WRAP) – an organisation that works with businesses and organisations to reduce food waste and curb plastic use – has published its latest report, which has been tracking progress in food waste reduction since 2007.

According to the report, titled *Courtauld Commitment 2025*, the UK is making significant steps in reducing food waste, with total levels falling by 480,000 tonnes between 2015 and 2018, the equivalent of seven per cent per person.

Similarly, the volume of food waste generated in the retail supply chain, the hospitality sector, and in homes stood at 9.5 million tonnes in 2018, down from 10 million tonnes in 2015 and 11.2 million in 2007.

WRAP suggested the decline could be linked to a number of factors including heightened public awareness, clearer labelling on food packaging and more local authorities offering food waste collections.

Despite the progress, the report details that UK households are continuing to waste 4.5m tonnes of food annually, which it estimates to be worth £14 billion – the equivalent to £700 a year for an average family with children.

WRAP adds that household food waste represents 70 per cent of all food waste triggered after the food has been grown or produced, with potatoes the single most wasted food.

Marcus Gover, chief executive of WRAP, said: 'We are in a new decade and have just 10 years if we are to honour our international commitment to halve food waste. This really matters because it is untenable that we carry on wasting food on such a monumental scale when we are seeing the visible effects of climate change every day, and when nearly a billion people go hungry every day.

'This means we are starting to wake up to the reality of food waste, but we are too often turning a blind eye to what is happening in our homes. We are all thinking about what we can do for the environment and this is one of the most simple and powerful ways we can play our part. By wasting less food, we are helping to tackle the biggest challenges this century – feeding the world while protecting our planet.'

Theresa Villiers, environment secretary, said the results were 'encouraging' but that more needed to be done.

'I urge all households, individuals and businesses to consider how they can reduce their own food waste footprint to create a better world for generations to come,' she said.

The reported also stated that supermarket food waste has risen slightly to 277,000 tonnes in 2018, compared to 260,000 three years earlier.

WRAP suggested the increase could be linked to efforts to help suppliers and customers cut waste, explaining that these can, in the short-term, increase food waste in depots and stores.

Last year, more than 100 UK food businesses and organisations pledged to reduce their food waste following a call to action from the government.

Tesco, Marks & Spencer, Waitrose and Unilever were among the companies that vowed to halve food waste by 2030 in the UK.

Judith Batchelar, director of Sainsbury's Brand, described food waste as 'one of the biggest challenges currently facing today's society', adding that it is 'an intrinsic part of our combined response to tackle greenhouse gas emissions and climate change'.

The retailer has already introduced various measures to reduce food waste, such as working with farmers to ensure that the amount of produce they provide is in line with quantities the supermarket's customers will buy.

24 January 2020

Towards more sustainable diets

Over 9 billion people are foreseen on Earth by 2050. With an increasing population comes a growing demand for food. In addition, many of us, especially in wealthy regions, consume more food than we need, and our diets are rich in animal-based products, a combination that has major negative impacts on the environment. To be able to provide enough food for future generations, while minimising our environmental impact, we must switch to more sustainable food production and change our dietary patterns.

The United Nations Food and Agriculture Organization (FAO) defines sustainable diets as having a low environmental impact, while meeting current nutritional guidelines, all while remaining affordable, accessible and culturally acceptable. But what changes do we need to make to achieve this vision?

The environmental impact of food production

The current food production system is recognised as one of the main causes of environmental damage, including climate change and the loss of natural resources. Agriculture alone is responsible for up to 30% of man-made greenhouse gas emissions (GHGEs), and 70% of our water use. It is the leading cause of deforestation, land use change, biodiversity loss, and a major source of water pollution and freshwater consumption. Other activities involved in food production and consumption, including raising livestock, transport, packaging and food waste, are also responsible for the environmental impact. Achieving a sustainable food production system and reducing food losses and waste are important global challenges that can help tackle the increased demand for food and help us to sustainably produce enough nutritious food for all.

The three pillars of a sustainable diet

Which steps to take as individuals wishing to achieve a healthy and sustainable diet might seem confusing at first. Despite the complexity, there are three changes that we can all make to achieve a more sustainable diet: consume less, waste less, and reduce our consumption of animal products in favour of plant-based alternatives.

Consume less

There is a global trend towards overconsumption, despite many around the world remaining hungry. While overconsumption has historically been a problem of developed countries, it is now a major issue in the developing world, particularly in emerging economies like China and Brazil. Overconsumption contributes to an increase of overweight and obesity, while at the same time driving unnecessary demand for increased production of crops and livestock with the associated environmental impact. Lowering our overall energy intake, especially in high-consuming countries, can benefit the health of both the environment and the population.

Waste less

In Europe, an estimated 88 million tonnes of food is discarded every year. Food is wasted during all stages of the food chain, by producers, processors, retailers, and caterers. However, most food waste, about 53% in Europe, takes place at home. Producing food that is then thrown away represents an unnecessary waste of land, water, labour and energy, and futile contribution to GHGEs. If food waste was a country, it would be the third largest producer of CO_2, trailing only the USA and China!

Less animal-based, more plant-based

In general, producing animal-based foods is more resource-intensive than plant-based foods, and has a higher environmental impact (such as land use, fresh water consumption and CO_2 emissions per tonne of protein consumed). Animal-based products also differ in their environment al impact. For instance, 7 times more CO_2 is produced to make 1 kg of beef compared to the same quantity of chicken. Insects, a common food source outside Europe, are also emerging as a protein source, with the potential to produce less GHGEs, and use less resources than conventional animal agriculture for similar amounts of protein. Choosing more sustainable animal-based products such as poultry, sustainably-grown fish or insects, reducing the consumption of animal-based products such as meat, dairy and eggs in general, and incorporating more plant-based products, such as fruits, vegetables, grains and legumes, are great steps towards a more sustainable diet. Diets high in plant-based foods have also been linked to a decreased risk of hypertension, stroke, type 2 diabetes, and certain cancers.

For those of us that choose to eliminate animal products from our diets entirely, a vegetarian or vegan diet can provide enough protein if the sources are diverse. A diversity of sources is important because some nutrients, including protein and essential amino acids, are found in smaller quantities in plant-based foods when compared to meat or fish. Therefore, what one source lacks can be compensated by another. The EU-funded project Protein2Food is working towards creating innovative plant-based products with enhanced protein content and quality. In the future, these products could represent an attractive option of protein-rich food for people who are willing to incorporate more plant-based food to achieve a more sustainable but equally nutritionally balanced diet.

Sustainable diets: not an easy concept

Achieving a sustainable diet may require trade-offs. For instance, buying locally-grown food may seem a sustainable choice, but only when the products are in season where we live. This is because the energy consumed to grow fruits and vegetables in heated greenhouses in winter is much higher than that needed to transport them from warmer countries.

Similarly, environmental and health benefits may sometimes not go hand in hand. For example, the benefits of fish consumption, particularly due to their omega-3 content, are well documented. However, over fishing and depletion of some fish stocks is already a problem, and if we all increased our fish consumption in line with dietary guidelines, the situation could get even worse. Research is being done into the development of oilseeds containing higher levels of omega-3's, and omega-3 enriched chicken is already a reality in supermarkets. These innovative products could help us meet our nutritional needs without putting pressure on our oceans. In the meantime, choosing to consume fish that is labelled as 'sustainably sourced' is a choice we can all make.

Small steps to achieve big goals

Whatever the sustainable diet approach followed, changes need to be realistic. Even small changes, on a global scale, can have a huge effect in reducing the environmental impact of food consumption. For instance, following a vegetarian or flexitarian diet (i.e. deliberate reduction of animal products consumption by substitution with plant-based alternatives), substituting ruminant meat (e.g. beef) with pork, poultry, or insects that have a lower ecological footprint, and choosing sustainably sourced fish and seafood, are small steps that nonetheless can produce huge effects on the global environmental impact of our diets.

19 April 2018

9 practical tips for a healthy and sustainable diet

Each of us contributes to the impact that our food system has on the planet. We can all commit to making the world a healthier place to live, through small but achievable changes to our diets.

1. Eat more fruits and vegetables

Fruit and vegetables are good for our health, and most come with a low environmental impact. There are exceptions, as some require a lot of resources to transport and keep fresh, so eating these less frequently can increase the sustainability of our diets. Examples include:

◆ fruits and vegetables that are fragile, or require refrigeration (salads and berries)

◆ vegetables that are grown in protected conditions (such as hot-house tomatoes or cucumbers)

◆ foods that use a lot of resources during transport (green beans, mange-touts, or berries imported from the southern hemisphere).

2. Eat locally, when in season

Locally-grown foods can be a sustainable choice, if we choose those that are in season where we live. The cost of producing or storing local foods beyond their natural growing seasons could be higher than shipping foods that are in season somewhere else.

3. Avoid eating more than needed, especially treats

Consuming only what we need reduces demands on our food supply by decreasing excess production. It also helps to keep us healthy and avoid excessive weight gain. Limiting snacking on energy-dense low-nutrient foods and paying attention to portion sizes are all useful ways to avoid unnecessary overconsumption.

4. Swap animal proteins for plant-based ones

In general, more resources are needed to produce animal-based proteins (especially beef), compared to plant-based proteins (such as beans, pulses and some grains). Eating a more plant-based diet also brings health benefits: plant-based food provides more fibre, and has a lower saturated fat content, both of which can contribute to a decreased risk of cardiovascular disease.

◆ For meat-eaters, limiting meat consumption to 1-2 times a week, having meat-free days and choosing more sustainable meats like chicken over beef can help us reduce our ecological footprint.

◆ For those choosing a vegan/vegetarian diet, combining different sources of plant-based protein will ensure our protein needs are met.

5. Choose wholegrains

Non-refined cereals are generally less resource intensive to produce than refined ones as they require fewer processing steps. They are also good for health, reducing our risk of cardiovascular diseases, type 2 diabetes, and overweight.

◆ Wholemeal bread, wholegrain pasta, unrefined barley, buckwheat and quinoa, are great choices.

◆ Brown rice is a good substitute for white rice, but it should be enjoyed in moderation, as a lot of water is used during its production.

9 practical tips for a healthy & sustainable diet

1 Eat more fruits & vegetables

2 Eat locally, when in season

3 Avoid eating more than needed, especially treats

4 Swap animal protein for plant-based ones

5 Choose wholegrains

6 Choose sustainably sourced seafood

7 Eat dairy products in moderation

8 Avoid unnecessary packaging

9 Drink tap water

Source: EUFIC

6. Choose sustainably sourced seafood

Fish is a good source of healthy omega-3 fatty acids, which contribute to normal vision, brain function and heart health. However, overfishing is causing wild fish stocks to become depleted. In order to benefit from the necessary nutrients and reduce pressure on wild fish stocks:

- consume fish and seafood 1-2 times weekly to provide the necessary nutrients and reduce pressure on wild fish stocks.
- choose fish and seafood marked with a sustainability label from certified organisations such as the Marine Stewardship Council.

7. Eat dairy products in moderation

While milk and dairy production has an important environmental impact, dairy products are an important source of protein, calcium and essential amino acids, and have been linked to reduced risk of several chronic diseases, including metabolic syndrome, high blood pressure, stroke, bowel cancer and type 2 diabetes.

- Enjoy low-fat unsweetened dairy products daily, but in moderation.
- Limit consumption of high-fat cheeses to occasional.

- For those of us who choose to eliminate dairy completely, opt for plant-based drinks that are fortified with vitamins and minerals, like calcium.

8. Avoid unnecessary packaging

Food packaging, especially when made of non-recyclable materials can have a huge impact on the environment. We all can reduce the amount of packaged products we buy (think of bulk apples versus cling-film wrapped ones), or opt for materials that are biodegradable, fully recyclable, or made from recycled materials.

9. Drink tap water

In Europe, the standards of water quality and safety are high. Instead of buying bottled water, we can re-fill a reusable water bottle at the tap as many times as we want. Tap water costs a fraction of the price of bottled water and reduces our ecological footprint.

19 April 2018

Ugly veg: supermarkets aren't the biggest food wasters – you are

An article from *The Conversation*.

By Miriam C. Dobson, PhD Researcher in Urban Agriculture, University of Sheffield & Jill L. Edmondson, EPSRC Living with Environmental Change Research Fellow, University of Sheffield

'Ugly' or 'wonky' veg were blamed for up to 40% of wasted fruit and vegetables in 2013, as produce was discarded for failing to meet retailer appearance standards. About 1.3 billion tonnes of food is wasted worldwide every year and, of this, fruit and vegetables have the highest wastage rates of any food type. But just how much of that is due to 'ugly veg' being tossed by farms and supermarkets? The biggest culprit for food waste may be closer to home than we'd like to admit.

'Ugliness' is just one reason among many for why food is wasted at some point from farm to fork – there's also overproduction, improper storage and disease. But the problem of 'wonky veg' caught the public's attention.

A report published in 2017 suggested that sales of 'wonky veg' have risen in recent years as retailers have acknowledged the problem with wasting edible food, but it's estimated that up to 25% of apples, 20% of onions and 13% of potatoes grown in the UK are still wasted on cosmetic grounds.

Morrisons reported that consumers had begun to buy more misshapen food, whereas Sainsbury's and Tesco both report including 'wonky veg' in their recipe boxes, juices, smoothies and soups.

Not all ugly veg is wasted at the retail point of the supply chain however. WRAP, a charity who have been working with governments on food waste since 2000, have investigated food waste on farms and their initial findings suggest a major cause of fruit waste is due to produce failing aesthetic standards. For example, strawberries are often discarded if they're the wrong size for supermarkets.

The National Farmers Union also reported in 2014 that around 20% of Gala apples were being wasted prior to leaving the farm gate as they weren't at least 50% red in colour.

Home is where the waste is

Attitudes seem to be changing on 'ugly veg' at least. Morrisons ran a campaign to promote its 'ugly veg' produce aisle, and other supermarkets are stocking similar items. Despite this, household waste remains the biggest culprit for food waste in the UK. Just under 5 million tonnes of food wasted in the UK occurs in households – a staggering 70% of all post-farm gate food waste.

A further million tonnes is wasted in the hospitality sector, with the latest government report blaming overly generous portion sizes. This suggests that perhaps – despite the best effort of campaigns such as Love Food Hate Waste – farms and retailers have been unfairly targeted by the 'wonky veg' campaigns at the expense of focusing on where food waste really hits home. The 2013 *Global Food Security Report* put the figure for household and hospitality waste at 50% of total UK food waste.

There are some signs we're getting better at least. WRAP's 2015 research showed that, at the household level, people now waste 1 million tonnes of food per year less than they did in 2007. This is a staggering £3.4 billion per year saved simply by throwing less edible produce away.

As climate change and its influence on extreme weather intensifies, reducing waste from precious food harvests will only become more important. Knowing exactly where the majority of waste occurs, rather than focusing too much on 'wonky veg' in farms and supermarkets, is an important step towards making sure everyone has enough affordable and nutritious food to live on.

During the UK's 'Dig for Victory' campaign in World War II, a large proportion of the population had to grow their own fruit and vegetables. Now the majority of people live in cities and towns – typically detached from primary food production. In the UK, the MYHarvest project has started to uncover how much 'own-growing' contributes to the national diet and it seems demand for land to grow-your-own is increasing.

Research in Italy and Germany found that people who grow their own food waste the least. One way to fight food waste at home then – whether for 'wonky' fruit and vegetables or otherwise – may be to replace the farm-to-fork supply chain with a garden-to-plate approach.

13 March 2019

Insects are 'food of the future'

Research from the University of Leeds, in England, and the University of Veracruz, in Mexico, concludes shoppers have to get past the 'ick factor'.

By Amy Murphy

Edible insects could be a solution to avoiding a global food crisis if consumers can overcome barriers such as the 'ick factor', a new study has found.

Insects are an environmentally sustainable food source, with a significantly lower carbon footprint compared to meat production, the report published in the *Comprehensive Reviews in Food Science and Food Safety* journal revealed.

But attitudes towards eating insects and current farming techniques and technologies need to change if edible insects are to become a common food source, researchers from the University of Leeds and the University of Veracruz, in Mexico, said.

Farming

The report found that edible insect cultivation remains rare in Western countries, where eating insects is still considered unusual, while negative perceptions have taken root in some countries where insects have been eaten traditionally, with the younger population associating it with poverty.

The best way of normalising edible insects is to target the preferences of the younger generation, who showed interest in using edible insects in unrecognisable forms, such as in flour or powder used in cookies or energy drinks, the authors said.

Another successful strategy to making edible insects become part of the mainstream is to serve them as snacks between meals, the report found.

The study, *Edible Insects Processing: Traditional and Innovative Technologies*, was carried out as worldwide food security faces serious risks, including the rapidly changing climate and an expanding global population.

The authors reviewed current insect farming methods, processing technologies, commercialisation techniques and current perceptions towards entomophagy - the practice of eating insects.

Sustainable

They revealed that edible insects have a high nutritional value and are a viable option as a sustainable source of protein.

Insect farming, which can be carried out in urban areas, uses much smaller amounts of land, water and feed and produces far fewer greenhouse gases compared to meat production.

But, if edible insects are to become a common food source, more development is needed to make the leap from wild harvesting to large-scale indoor farming that could meet demand in an economically efficient, safe and sustainable manner, the study found.

Food

Dr Alan-Javier Hernandez-Alvarez, one of the study's authors, from the School of Food Science and Nutrition at the University of Leeds, said: 'Edible insects are fascinating. Although humans have eaten insects throughout history, and approximately two billion people around the globe regularly eat them today, research on the subject is relatively new.

'Edible insects could be the solution to the problem of how to meet the growing global demand for food in a sustainable ways.

'The "ick factor" remains one of the biggest barriers to edible insects becoming the norm. Eating behaviour is shaped largely during early childhood and in Western countries, eating insects, especially in whole and recognisable forms, remains something seen mostly on TV shows.'

Dr Guiomar Melgar-Lalanne, study author from the University of Veracruz, added: 'In Western countries it is the younger generation that show more willingness to try new food products, including edible insects.

'The "foodies boom" and the rise of veganism and flexitarians have opened the door to alternative food sources.

'Promoting insects as an environmentally sustainable protein source appeals to the current attitudes in the younger generation.'

She continued: 'But if edible insects are to become a common food source, current farming techniques and technologies could struggle with the demand and need to be expanded.'

2 July 2019

Recycling and refurbing old phones tackles our appetite for newer, shinier gadgets

The UK produces more e-waste than the EU average, much of which contains chemicals capable of contaminating water and soil.

By Rhiannon Williams

When you last upgraded your phone, what did you do with your old handset? If you left it in a drawer to gather dust, you're not alone.

Recent research by the Royal Society of Chemistry found that 51 per cent of UK households are holding onto at least one unused electronic device – the equivalent of around 40 million old mobile phones, laptops, tablets, e-readers, desktop PCs and TVs.

More of us are using technology than ever before – 79 per cent of UK adults personally use smartphones, according to Ofcom, while Ipsos Mori research suggests one in three own an iPad – but all this increased connectivity is not without consequence.

Our insatiable appetite for newer, shinier gadgets, falling prices and manufacturers designing products to break after just a few years (known as inbuilt-obsolescence) has spelt disaster for the planet. Around 50 million tonnes of it is produced each year (the equivalent of every commercial aircraft ever constructed), but only around 20 per cent is ever formally recycled, according to the World Economic Forum.

Dumping e-waste

The UK produces more e-waste than the EU average – 24.9kg of e-waste per person, compared to Europe's 17.7kg – much of which contains hazardous chemicals and substances capable of contaminating water and soil.

Dumping e-waste is also a missed opportunity to extract valuable metals and elements. Six elements currently used in phones (gallium, arsenic, silver, indium, yttrium and tantalum) could be fully mined within 100 years if more sustainable alternatives are not sought alongside greater efforts to reclaim them through recycling, Dr Mindy Dulai, senior policy adviser at the Royal Society of Chemistry, tells *i*.

'Tantalum, for example, is really important in making electronic components, while indium tin oxide is used in touchscreens, both of which are set to run out in the next 100 years,' she says. 'We've got to think about using these items for longer, or if they're functional give it away to someone, a charity maybe, or sell or recycle them.'

While efforts to recycle disposable coffee cups and single-use plastic have made headlines across the UK, it's not always transparent where your old smartphone ends up if you've dropped it off to be recycled. Chances are if you've used one of the UK's many websites dedicated to offering cash for old gadgets, including musicMagpie and Mazuma Mobile, it's ended up in an innocuous-looking building off the Kingston Bypass in Surbiton, where close to 100 staff are quietly beavering away to turn the tide on e-waste.

Recycling facility GSUK works with leading tech brands, including Sony, Microsoft, Samsung and Huawei, phone operators and recycling websites to recover, recycle, redistribute and reuse old electronic products to prevent them from piling up in landfill or polluting the environment through incineration.

Dominic Strauli, device recovery manager at the plant, joined GSUK a year ago from gadget repair company iSmash, coaching staff in how to both speed up how the items are processed while extracting intact components safely and efficiently.

Reselling overstock

Around 85 per cent of the devices it receives from its client are recycled into refurbished products, processing around 20,000 phones and 250,000-400,000 units of accessories – charging cables, plugs etc – each month.

'Some devices are much easier to reclaim than others. Many of the high-end phones from Apple and Samsung are made from metal and glass, they're very recyclable,' he tells *i*. 'Wearables, smartwatches and the styli you use with phones and tablets, they're much harder. Devices which use a lot of plastic too – plastics are very hard for a recycling plant because of the mess they make.'

Many of the devices GSUK processes are unsold stock from networks or manufacturers, or items customers have returned after briefly using. Its warehouse contains thousands of neatly-stacked cardboard boxes filled with bundles of

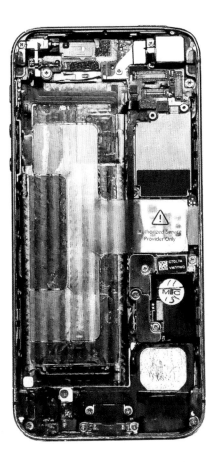

phones, tablets, smartwatches, headphones and cables, all of which are checked to ensure they turn on and whether they contain any customer data.

If present, the data is then wiped, the device checked for faults (broken screen, battery faults etc) and repaired for reselling. Its sister company, Genuine Solutions Limited, also distributes previously unwanted technology products to independent retailers, e-tailers, wholesalers to sell.

If a phone won't turn on or can't be wiped of its data, it's fully disassembled for the team to examine and test its components – screen, back cover, battery, housing, cameras etc – all of which can be sold on.

Plea to manufacturers

The motherboard (the brains of the device) is then held for destruction and metal extraction by GSUK's recycling partner, shredding it into tiny pieces before it's run through a fluid to extract its precious metals and gold into tanks.

GSUK has recovered more than 22.7 million products, some of which are sold into developing markets while others are distributed to people in need through a partnership with phone network Three. The second-hand electronics market is 'booming', Dominic claims, but warns consumer pester-power can only flourish amid change from the manufacturers as well as legislating governments.

'Manufacturers need to make devices that are much easier to be recycled or used or repaired. These big companies need to pull their socks up and realise there's an opportunity here to save the planet, because they need to do more,' he says.

'A few years ago, the average life cycle of a smartphone was 2.3 years. Now it's around 2.6 years, because people are starting to hold onto them for longer and then passing them on to friends and family, which is a great place to start. Once they decide to upgrade to a new one, recycling it ethically will, collectively, make a huge difference, and could make them money – which is good for the consumer and the environment.

'Before I worked here, I was never really into recycling. Now I live and breathe it. Radical change needs to happen, and it needs to happen now.'

6 November 2019

Key Facts

- 14% of all greenhouse gas emissions come from transport. (page 3)
- Millennials are far more likely to shop ethically, eat more locally-sourced food and adopt a vegan diet. (page 5)
- When asked if they would pay more for eco-friendly products, 61 per cent of millennials also say they would willingly do so, whilst only 46 per cent of baby boomers said likewise. (page 5)
- Millennials make up around 25 per cent of the world's population and 84 per cent of them believe it is their responsibility to change the world for the better. (page 5)
- The UK ranked 12th out of a possible 28th in Europe on target to meet the Sustainable Development Goals. (page 8)
- Salford has lower-than-average CO2 emissions, lower-than-average energy consumption levels and higher-than-average recycling rates. (page 9)
- Every year 27 million cars are wrecked and 25 million television sets are discarded. (page 10)
- An estimated 500 billion to 1 trillion plastic bags are consumed worldwide each year. (page 11)
- Over 182.5 billion plastic straws are used per year. (page 11)
- In Australia, 1 billion takeaway hot drink cups are thrown away every year. (page 11)
- (83%) of Britons say that more needs to be done to encourage recycling in the UK. (page 12)
- One in six Brits (16%) say they would recycle more if their local authority collected more types of recyclables, more often, from their kerbside bins. (page 12)
- 10% of Brits say they think they already recycle all of their waste, and nothing could help them to recycle more. (page 12)
- Household living costs in the UK typically rose by 8%-12% from 2015 to 2019. (page 16)
- If you eat a packed lunch every day at school you will send around 2,275 sandwich bags to landfill. (page 18)
- The UK generates around 222.9 million tonnes of waste each year. (page 19)
- In 2017 the total waste from households in England alone was 22.4 million tonnes, which is the equivalent of 403kg per person – roughly the weight of two motorbikes. (page 19)
- The #ZeroWaste hashtag has more than 2.6 million Instagram posts. (page 19)
- The plastic we use today will exist on the earth long after we're gone. (page 20)
- Making enough plastic water bottles for the United States requires 17 million barrels of oil and enough energy to power 190,000 homes for a year. (page 20)
- UK supermarkets are responsible for churning out at least 59 billion pieces of plastic every year. (page 20)

- Between 400,000 and one million people die every year because of diseases and accidents linked to poorly managed waste in developing countries. (page 22)
- Air pollution is responsible for 3.7 million premature deaths a year. (page 22)
- Of the 9 million tonnes of plastic that ends up in the ocean every year, around 2,000 tonnes of that is made up of plastic straws. (page 23)
- Nearly a trillion plastic bags are still used worldwide each year. That works out at 2 million every minute. (page 24)
- UK households are getting through 54 'bags for life' each year suggesting that they are being used as a bag for a week. (page 25)
- 50% of all sea turtles have ingested pieces of plastic. (page 25)
- At least 50,000 particles of microplastics make their way into our bodies each year. (page 25)
- The fashion industry is responsible for 20 per cent of global water pollution. (page 26)
- Up to 90 per cent of charity shop donations are sent to developing countries where most items go unwanted. (page 26)
- The fashion industry creates in excess of 80 billion pieces of clothing a year and is responsible for 10% of global carbon emissions. (page 28)
- Less than 1% of fabrics can be recycled. (page 28)
- On average, cheap dresses and other clothes products are discarded by consumers after five weeks. (page 30)
- The textile industry creates 1.2 billion tonnes of CO2 a year, more than international aviation and shipping combined. (page 30)
- Some garment workers in Leicester are paid an average of £3 an hour – way under the national minimum wage. (page 30)
- In the UK, we buy more clothes per person than any other country in Europe. (page 30)
- UK households are wasting 4.5m tonnes of food that could be eaten ever year. (page 32)
- Household food waste represents 70 per cent of all food waste. (page 32)
- The cost of household food waste is equivalent to £700 a year for an average family with children. (page 32)
- Agriculture is responsible for up to 30% of man-made greenhouse gas emissions and 70% of our water use. (page 33)
- In Europe, an estimated 88 million tonnes of food is discarded every year. (page 33)
- 'Ugly' or 'wonky' veg were blamed for up to 40% of wasted fruit and vegetables in 2013. (page 36)
- Up to 25% of apples, 20% of onions and 13% of potatoes grown in the UK are wasted on cosmetic grounds. (page 36)
- The UK produces more e-waste than the EU average – 24.9kg of e-waste per person, compared to Europe's 17.7kg (page 38)

Biodegradable waste

Materials that can be completely broken down naturally (e.g. by bacteria) in a reasonable amount of time. This includes organic materials such as food waste, paper waste and manure, which can be composted, as opposed to items such as plastic bottles that would take thousands of years to break down naturally.

E-waste

Electronic waste; discarded electrical items such as mobile phones and computers. There are strict EU regulations in place to ensure that e-waste is safely recycled or disposed of: however, the shipping of e-waste to developing countries is becoming an increasingly common problem.

Eco-friendly

Policies, procedures, laws, goods or services that have a minimal or reduced impact on the environment.

Ethical consumerism

Buying things that are produced ethically – typically, things which do not involve harm to or exploitation of humans, animals or the environment; and also by refusing to buy products or services not made under these principles.

Fairtrade

A movement which advocates fair prices, improved working conditions and better trade terms for producers in developing countries. Exports from developing countries that have been certified Fairtrade – which include products such as coffee, tea, honey, cocoa, chocolate, sugar, cotton and bananas – carry the Fairtrade mark.

Fast fashion

Inexpensive, mass-produced clothing that is usually produced quickly to respond to current fashion trends. Often, it is only worn a few times before being thrown away.

Food waste

Around seven million tonnes of food is thrown away by households in the UK every year. Some of the waste is unavoidable, such as peelings or bones, but most of the food is edible. This is because there is often confusion over use-by and best-before dates. Also, many families buy more food than they actually need..

Global footprint

A person's global footprint refers to the impact that they have on the planet and the people around them, taking into account how much land and water each person needs to sustain their lifestyle.

Green energy

The same as renewable energy, which comes from natural resources rather than non-renewable sources. It's called 'green' due to the fact that the sources are environmentally friendly, sustainable and have zero emissions.

Greenwashing

'Greenwashing' occurs when organisations falsely promote or market themselves as having 'green', environmentally friendly, practices.

Infrastructure

The basic, interrelated systems and services needed to underpin a community or society, such as transport and the provision of power and communication systems, as well as public institutions such as schools.

Pollution

Toxic substances which are released into the environment: for example, harmful gases or chemicals deposited into the atmosphere or oceans. They can have a severe negative impact on the local environment, and in large quantities, on a global scale.

Recycling

The process of turning waste into a new product. Recycling reduces the consumption of natural resources, saves energy and reduces the amount of waste sent to landfills.

Resource consumption

The use of the Earth's natural supplies, including fossil fuels, water, wood, metals, minerals and many others. Growing populations and increased standards of living have resulted in increased consumption of natural resources, which is having a negative effect on the environment.

Sustainability

Sustainability means living within the limits of the planet's resources to meet humanity's present-day needs without compromising those of future generations. Sustainable living should maintain a balanced and healthy environment.

Sustainable

Something that is capable of being maintained at a particular rate or level.

Sustainable Development Goals (SDGs)

17 goals set out by the United Nations to protect the planet and ensure that people around the world can live with equality and in a healthy environment by 2030. The goals cover social, economic and environmental sustainability.

Sustainable diet

Sustainable diets have a low environmental impact – this includes the impact of food production and consumption on our planet's resources

Waste

Anything that is no longer of use and thrown away. Each year the UK generates approximately 290 million tonnes of waste, which has a damaging effect on the environment.

Activities

Brainstorming

- In small groups, brainstorm to find out what you know about sustainability. Consider these questions:
  - What do we mean by 'sustainability'?
  - Why is achieving a sustainable lifestyle important?
  - What is sustainable development?
  - What is fast fashion?
  - How can we prevent waste?
- In pairs, think about the R's of sustainability. Usually these are reduce, reuse and recycle. How many more can you list?
- Create a Diamond 9 of the issues that affect sustainability, with the most important at the top and least important at the bottom. Discuss with a partner why you have listed them this way.

Research

- Using this book, and the Internet, research the Sustainable Development Goals set by the UN. Choose one of the goals and research why it has been set and what will need to happen in order to achieve it, and make some notes.
- Research sustainability schemes in your area. Choose your favourite idea and create a presentation that will showcase it to the rest of your class.
- Choose a company and research their policies on sustainability, make notes and share with a partner.
- Research charities that promote sustainability. Make a list of those charities and which section of sustainability they are working towards.
- Research how much food is wasted each day in the UK, and how this can be prevented.
- Using the article *Insects are 'food of the future'*, create a questionnaire to find out which insects your friends and family would be willing to eat. Consider how they could be presented and whether people would prefer eating the insect whole, or in a product such as flour. Present your findings in a graph.
- Using the article *Britons say more needs to be done to encourage recycling* as inspiration, create a short survey to find out your classmates ideas on recycling and what could be done to promote it.

Design

- Choose one of the articles from this book and create an illustration to accompany it.
- In small groups, imagine you work for a charity that promotes sustainable lifestyles among people in the UK. Design a leaflet to promote your charity's sustainable ethos.
- Design a poster to promote 'Make do and mend' to persuade people to reuse, recycle or repair their clothes.
- Design a leaflet to encourage people to waste less food.
- Choose one of the articles in this book and create an infogram to show the information.
- Design an app to promote slow fashion.
- Look at some of the packaging of food in your home. Why do you think it was designed this way? Is there any way you could improve the design? What materials could you use?

Oral

- In small groups, discuss methods to make your school more sustainable. Make a list of things that could be done.
- In pairs, talk about things that you already do that are sustainable. Now think of areas where you can improve, and what can be done to make your life more sustainable.
- As a class, debate the statement: 'Companies use sustainability to sell more products'.
- In small groups, discuss sustainable fashion. Can it ever be truly sustainable?
- Talk to your parents or grandparents about whether or not they used to keep products for longer than we do now. How did they do this? Did they repair items or throw them away? Why did they do this? Were they aware of the environmental impact or was it to do with cost?

Reading/writing

- Choose one of the articles about sustainable solutions and research it further. Write a blog post about your chosen solution.
- Over the course of a week, keep a diary detailing the resources that you use (water, electricity, wood, food, and anything else you can think of). At the end of the week, look back at your diary and think about whether there are any areas where you could cut down the resources you use. Write a summary explaining what you could change in order to achieve a more sustainable lifestyle.
- Write a letter to your head teacher suggesting some changes that your school could make in order to be more sustainable.
- Read the article *'I live a zero-waste life - here are my tips to reduce your impact on the planet'* and write down how you could reduce the amount of waste you make each day.
- Read the article *Sustainable living for beginners – a starter guide*, and write a short summary and pick out five key points.
- Watch the documentary *Drowning In Plastic*. Write down some key points from the film and write a short review.
- Write a short blog post on how you will reduce food waste in your home. Include things like hints and tips and recipes.
- Chose a supermarket and write a letter to them to persuade them to use less plastic. Include ideas for alternatives that they could use instead.

Acknowledgements

The publisher is grateful for permission to reproduce the material in this book. While every care has been taken to trace and acknowledge copyright, the publisher tenders its apology for any accidental infringement or where copyright has proved untraceable. The publisher would be pleased to come to a suitable arrangement in any such case with the rightful owner.

The material reproduced in *ISSUES* books is provided as an educational resource only. The views, opinions and information contained within reprinted material in *ISSUES* books do not necessarily represent those of Independence Educational Publishers and its employees.

Images

Cover image courtesy of iStock. All other images courtesy of Pixabay and Unsplash, except pages 1, 4, 10, 24: Freepik, 11: Rawpixel and 37: iStock

Icons

Icons on pages 25 were made by Freepik, Macrovector & Vectorpocket, 30 were made by Freepik, Eucalyp, Darius Dan, Srip, monkik & dDara, 35 were ma made by Freepik & Macrovector from www.flaticon.com.

Illustrations

Don Hatcher: pages 5 & 19. Simon Kneebone: pages 6 & 20. Angelo Madrid: pages 3 & 15.

Additional acknowledgements

With thanks to the Independence team: Shelley Baldry, Danielle Lobban, Jackie Staines and Jan Sunderland.

Tracy Biram

Cambridge, May 2020